Bureau of Land Management, and Utah
. Lehi Hintze of Brigham Young University
M. Bartley of the University of Utah were
, contributing their knowledge of Utah as
e draft manuscript. Don Hyndman and David
rsity of Montana at Missoula also read and
nuscript. Ray Strauss contributed many excel-
s of geological features. And Luiz Schleiniger
y negatives to bring resulting prints up to
ndards. I thank all of them most sincerely.

s book are adaptations of the Geologic Map of
d in 1980 by the Utah Geological and Mineral

ROADSIDE GEOLOGY of Utah

Halka Chronic

MOUNTAIN PRESS PUBLISHING COMPANY
Missoula, Montana, 1990

GREAT EVENTS IN UTAH

ERA	PERIOD	EPOCH	AGE	EVENTS
CENOZOIC	QUATERNARY *Q*	Recent	.01	Erosion trenches Pleistocene lake and stream deposits, provides fill for intermountain basins. Small eruptions create lava flows, cinder cones. Continued fault movement cuts alluvial fans and Lake Bonneville deltas.
		Pleistocene	2	Colder, wetter climates of glacial periods bring glaciation to mountains, development of Lake Bonneville and its ancient shorelines. Lava flows erupt in southwest Utah.
	TERTIARY *T*	Pliocene	5	Basin and Range faulting, canyon-cutting, and localized volcanism continue.
		Miocene	24	Block and detachment faulting begin, widening Basin and Range region. Hurricane, Wasatch, and Sevier faults develop. Basalt volcanism begins. Utah and surrounding states rise 5000 feet; with uplift, Colorado River and tributaries deepen their canyons.
		Oligocene	37	Scattered explosive volcanism covers much of Utah with volcanic ash. Small intrusions push upward, some rich with copper, iron, and other ores.
		Eocene	58	As mountains wear down, lakes between them continue to receive sediments, including oil shale.
		Paleocene	66	Uinta Mountains rise; lakes develop between them and the Colorado-Wyoming Rockies. Fine lake siltstones and silty limestones are deposited.
MESOZOIC	CRETACEOUS *K*		144	Last invading sea comes from east and southeast, brings near-shore sandstone and coal deposits, deeper-water shales. Sevier Mountains rise in western Utah, shed coarse sediments. Thrust faulting shoves Paleozoic sedimentary rocks eastward over younger rocks, narrowing Utah.
	JURASSIC *J*		208	Desert conditions extend over much of Utah, bringing dune sand, salt, and gypsum. Granite intrusions form in western mountains. Dinosaurs leave footprints and skeletons in sandstone-shale sequences.
	TRIASSIC *TR*		245	Red silt and mud, washed from Uncompahgre Highland, are deposited on tide flats. Then land rises; sea retreats westward. Volcanoes spew volcanic ash, which preserves many fossil trees.
PALEOZOIC	PERMIAN *P*		286	Fossil-bearing marine limestones are locally succeeded by fine dune sandstones. Uncompahgre Highland rises in eastern Utah and Colorado.
	PENNSYLVANIAN *IP*		330	Salt, gypsum, potash deposited in subsiding basin in southeast Utah. In Oquirrh Basin of northwest Utah, thick marine limestones and shales accumulate. Animal life includes brachiopods, corals, bryozoans.
	MISSISSIPPIAN *M*		360	Thick marine limestones are deposited, containing fossil corals, snails, brachiopods, bryozoans.
	DEVONIAN *D*		408	Most of state is still covered by sea, with resulting limestone, shale, sandstone. Uplift in north central Utah, where erosion removes some earlier strata.
	SILURIAN *S*		438	Dolomite is deposited over much of state. Fossils include corals, brachiopods, signaling a warm tropical sea.
	ORDOVICIAN *O*		505	Marine limestone, sandstone, dolomite are deposited in that order, thickest in western Utah. Trilobites, brachiopods, and other marine invertebrates are preserved as fossils.
	CAMBRIAN *€*		570	Sea advances across western continental shelf. Sandstone, then shale, then limestone accumulate as sea deepens. Invertebrate shellfish leave abundant fossils.
PRECAMBRIAN *P€*				**Younger:** Great thicknesses of sedimentary rocks accumulate, are later altered to marble, slate, and quartzite. Glaciation brings coarse gravel, later altered to tillite. A long period of erosion follows. **Older:** Major episodes of mountain building altered all pre-existing rock to gneiss and schist.

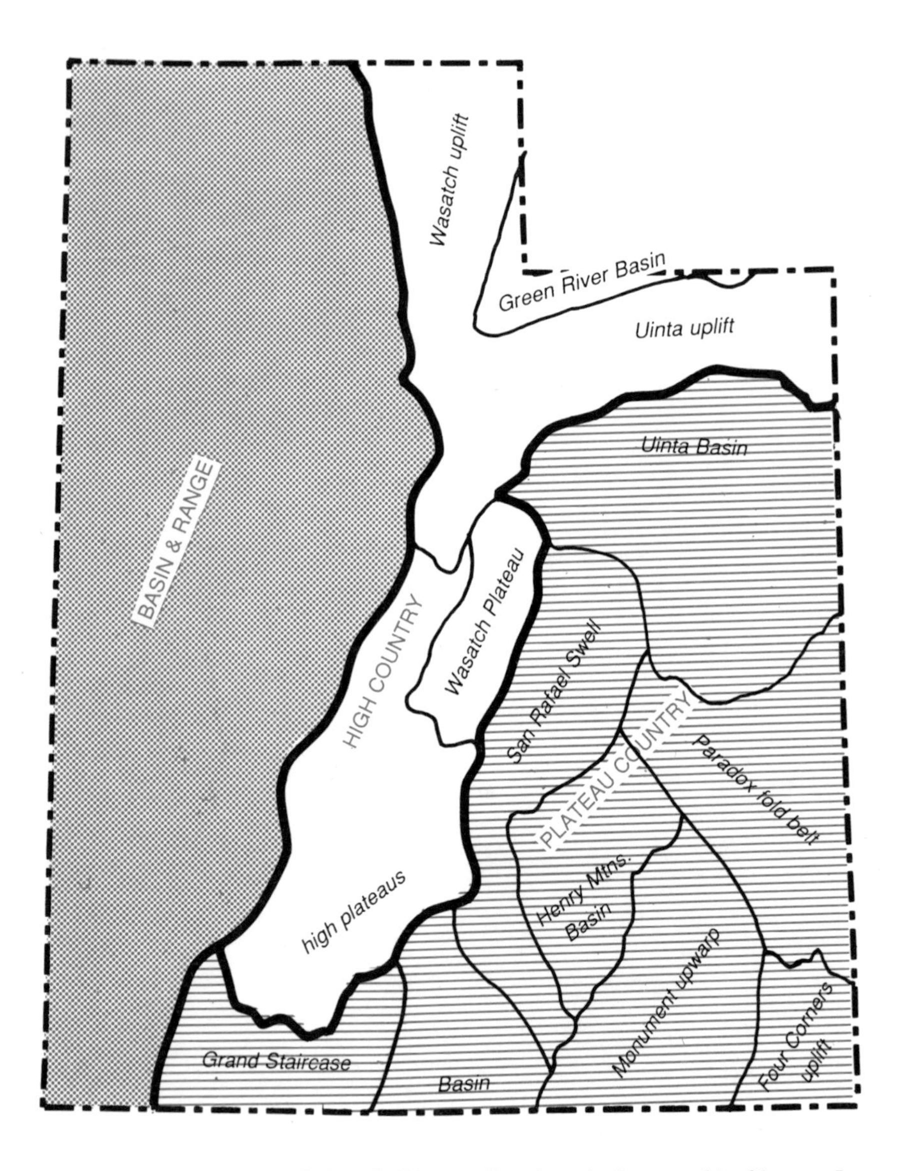

In this book, most of the Colorado Plateau Province is discussed in Chapter I. The High Plateaus, Wasatch Plateau, and Uinta Mountains are discussed in Chapter II. The Basin and Range region is described in Chapter III. Descriptions of National Parks and Monuments are in Chapter IV.

Contents

Rainbow Bridge, carved by wind and water, raises its massive arch above the creek that gave it birth. —Ray Strauss photo.

Getting Started

THE FACE OF THE EARTH

Utah's magnificent scenery stems from its geology. Mountains and mesas, rivers, lakes, and desert ranges — in as great a variety as you'll find anywhere — all have their geologic foundations. Here as elsewhere, rocks control the scenery, control the landscape, control even the positioning of towns and cities and the routing of highways and byways. So to understand Utah we must look first at basic geologic knowledge.

Geology as a science is younger than chemistry, physics, astronomy, and biology. Simply defined as the study of the Earth, it came into being for utilitarian purposes — to chart expected locations of coal seams and mineral deposits. For a long time, geology was a study just of the Earth's crust. Only in about the last 100 years have we begun to decipher the Earth's inner workings, its age in terms of years, the story of its oldest rocks. Only within the last 30 years have we recognized and begun to understand the wandering of its continents, the shifting of its poles, the unfolding of its lands and seas, its lowland plains, its mountains. To that knowledge something new is added every year, indeed every day, by geologists around the world, still seeking to decipher more precisely the nature and history of the planet we live on.

LOOKING DEEPER

Our Universe was born about 15 billion years ago in an immense explosion, the Big Bang. In the instant of the explosion, the material of all the stars and suns and planets and moons shot out in all directions. Where blobs of this material happened to collide, they were held together by gravity. Gradually, over several billion years, blobs grew larger and took on spherical shapes. The largest whirling balls of incandescent gas, heated by their own fusion reactions, became stars.

In widely scattered parts of the universe, great numbers of blobs found each other and whirled into galaxies. Though gravity would pull them together, individual blobs were held apart by their own whirling motions. In one particular galaxy, one particular star, our Sun, collected an array of smaller blobs that, whirling around it, cooled and became planets.

Our Earth cooled enough to become a planet about 4.6 billion years ago. By that time, it had changed from a ball of gaseous material into a ball of molten rock, and then into a relatively rigid mass of solids and semi-liquids, with a thin but solid crust.

As the inside of the planet cooled, its heaviest components sank toward its center, while the lightest rose to its surface. Geologists have learned from studies of the arrival times of earthquake waves that the Earth's center is a core of nickel and iron about 4,400 miles in diameter. Outside the core is the mantle, a seething layer 1800 miles thick, composed mostly of peridotite, a dense black rock especially rich in iron and magnesium. The lower part of the mantle is thought to be entirely solid, but its outer part contains a partially molten zone that deforms plastically, the way red-hot iron is plastic enough to be bent and shaped on a blacksmith's anvil. No blacksmiths work down there in the mantle, but there is stupendous pressure exerted by gravity, by the weight of the Earth's outermost layers. And there is heat, lots of it, generated by decay of radioactive minerals.

The Earth's solid crust floats, so to speak, on the mantle. The lower part of the crust and the upper 40 miles or so of the mantle seem to be coupled, to function together. Geologists now call this crust-upper mantle combination the lithosphere. Relative to the size of the Earth, the lithosphere is a pretty thin skin, about 40 miles thin under the oceans and 60 to 100 miles thin under the continents — as thin relative to the size of the Earth as the fragile film that forms on a bowl of hot chicken soup.

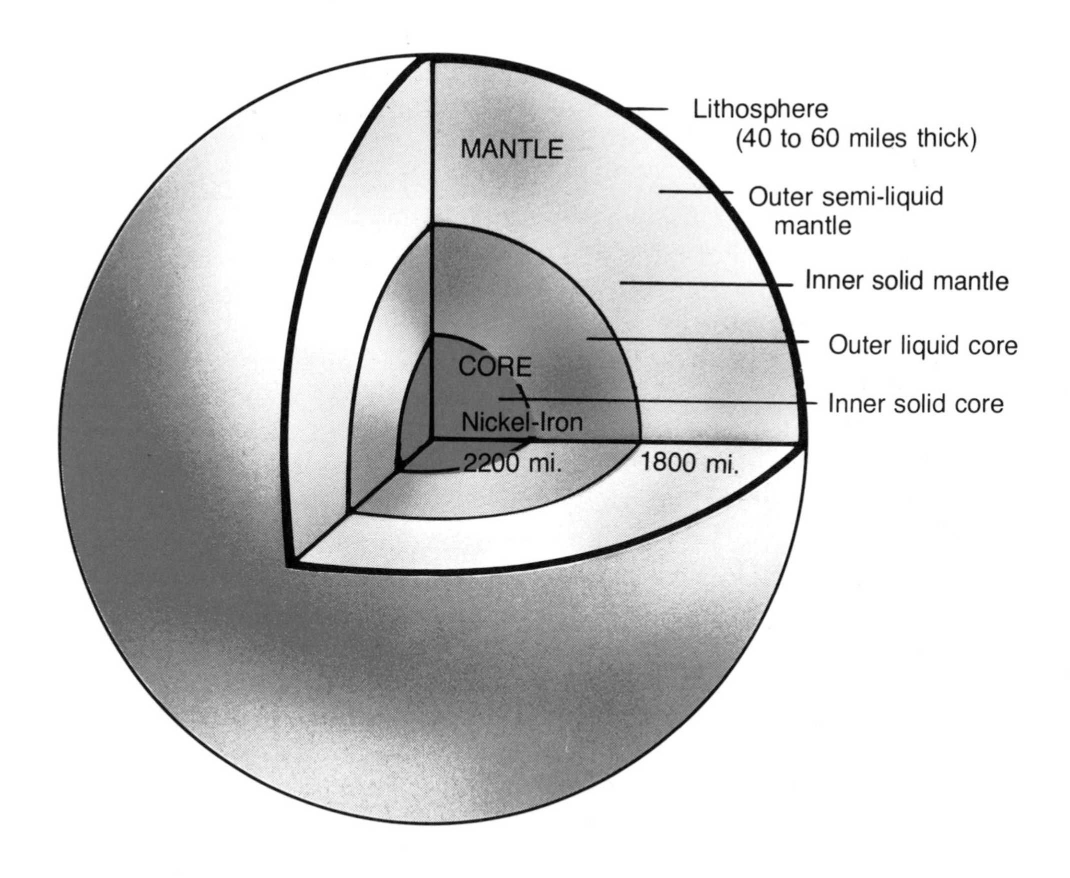

And speaking of soup, the extremely hot, semifluid upper part of the mantle seems to behave the way soup and other fluid substances behave when they are heated. It develops convection currents that rise and roll over and plunge downward again like the big "boils" in the soup kettle. But of course on a huge scale, and extremely slowly by human standards— moving at most only a couple of inches a year.

With the boils as they move horizontally at the top of their roll, the convection currents carry the film of the lithosphere on the surface. Still moving extremely slowly, they gradually tear it apart into large, fairly rigid plates. The currents shove these plates here and there, jostling them, pushing them upward or pulling them downward, bending and breaking the rocky layers, constantly reshaping the face of the Earth. All told, this movement and breaking has now created a dozen large lithosphere plates, as well as a number of smaller ones. Their movements are responsible for most of the geology and geography we know today.

Under the oceans the lithosphere consists of dark, heavy basalt of the oceanic crust floating on top of yet heavier, coarse-grained rock called peridotite, part of the upper mantle. Basalt is not quite as rich in heavy minerals as the mantle, and therefore is somewhat less dense. The lithosphere of the continents, on the other hand, includes lighter colored, lower density rocks on top of the heavy peridotite of the upper mantle. Most of these rocks are much richer in silica than either basalt or peridotite. Because of their lesser density, the continental crust floats higher than oceanic lithosphere — like foam on our kettle of soup. This is why our world has "the mighty oceans, and the pleasant land."

Many plates are edged by oceanic ridges or deep oceanic trenches. And there is a reason for this. Oceanic ridges are fractures along which the slowly rising rocks in the upper mantle reach nearly to the surface. The upward motion of mantle material pushes two adjacent oceanic plates apart, allowing molten rock, magma, to well up from the mantle and erupt onto the sea floor. This molten rock cools and solidifies to form bands of new crust along the crests of the oceanic ridges — at the same time pushing older crust aside in each direction. Oceanic trenches are lines where the opposite edges of the plates collide with adjacent plates, where one plate sinks beneath the other. If one plate is continental and the other oceanic, the denser oceanic plate tends to sink under the less dense continental plate, moving downward to melt and once more become part of the mantle. Oceanic trenches lie where the doomed plate begins to sink.

But plate boundaries do not generally coincide with the boundaries of continents. Many of the large plates, our own North American plate among them, are partly oceanic, partly continental. The oceanic part behaves like an oceanic plate, with periodic additions along an oceanic ridge, in our case the mid-Atlantic ridge. The continental part of the plate, propelled, like the oceanic part, by slow movements in the mantle, eventually collides with an adjacent oceanic plate — in our case a number of small plates that we can lump together as the East Pacific plate. The lighter continental part of the North American plate overrides the heavier, denser oceanic crust of the East Pacific plate.

But not without considerable damage. As in all collisions, there is a certain amount of fender-bending: Mountains are pushed up, parts of the continent are drawn down into the mantle, islands are tacked onto the continental mass. Earthquakes are frequent. Basalt magma may rise into the overlying continental crust. In some cases its heat melts part of the continental crust. Oceanic and continental types of magma may push upward to core long mountain ranges, or breaking through to the surface may build tall volcanoes along the continent's edge.

Utah lies on the western part of the North American plate, which reaches from the mid-Atlantic ridge to the Pacific coast. Despite its inland location away from the violence of the plate margins, Utah has seen a great deal of mountain building. Its mountains are formed by compression, tension, or by simple vertical uplift.

GEOLOGIC TIME

When geologists talk about Earth history, they speak in terms of time: When did this rock form? What relationship does it have with rocks formed earlier or later? Concepts of geologic time are hard to grasp, mostly because the Earth and its rocks are so much older than Man's brief sojourn on our planet. Geologic time, which began when the first rocks of the Earth's crust had cooled enough to be called rocks, involves several billion years.

Although they knew no way to tell exactly how old rocks were, early geologists could often figure out which of two rocks was the younger and which the older. They soon realized that sedimentary rocks, those deposited as layers of sediment in seas and lakes and on the continents, accumulate with the oldest layers at the bottom of the pile and the youngest on top. Geologists soon devised ways of telling whether sedimentary rocks are in their normal position, or overturned. As knowledge grew, a known sequence developed, a geologic calendar, quite without exact dates, but a calendar that could be and has been used by geologists everywhere ever since.

On this calendar the largest intervals — let's liken them to years — were called eras. Subdivisions of eras, which we can liken to months, were called periods. And subdivisions of periods, weeks if you like, were called epochs. Unlike years or months or weeks, the eras, periods, and epochs turned out to be quite uneven in length.

Eras were named according to the development and evolution of life: Paleozoic (ancient life), Mesozoic (middle life), and Cenozoic (recent life). The oldest rocks were recognized as being older than Paleozoic, and were called Precambrian — older than the Cambrian rocks.

Most periods were named for locations in which their rocks were first studied or were especially abundant: Cambrian for the old Roman name for Wales, Devonian for Devon in England, Jurassic for the Jura region of France, Permian for the province of Perm in Russia. There are some exceptions: Cretaceous was named with a Latin word for chalk, for the chalky white cliffs near Dover, England. Triassic got its name because in Germany, rocks of that period were easily divided into three parts. At first, rocks were assigned to these periods by study of their fossils, which showed the time-related evolution of life.

Era and period names are used fairly consistently over the whole world, with only minor differences from continent to continent. Epochs, on the other hand, vary from one continent to another. In this book, emphasis is on eras and periods, and the only epoch names used are North American terms for subdivisions of the Cenozoic era, as shown on the geologic calendar on page v.

ROCKS AND MINERALS

The Earth's crust, as we all know, is made of rocks such as sandstone and shale, granite and basalt, and others. And rocks are made of minerals — quartz, feldspar, mica, garnet, gypsum, salt, and so forth. Rocks can be classed by their origin. They fall into three categories: igneous, sedimentary, and metamorphic. All three categories occur in Utah, so let's look at each separately.

Much of the volcanic rock in Utah is breccia, formed of broken blocks of silicic lava with or without a matrix of fine volcanic ash.

Igneous rocks originate from molten rock, magma. When magma erupts on the Earth's surface, it becomes volcanic rock. Volcanic rock includes both lava and tuff, the latter formed of the fine airborne particles we know as volcanic ash. It can further be classified by its chemistry: Silicic volcanic rock is high in silica, light in color. Most silicic lavas are thick and sticky, but some contain enough volcanic gases such as steam to make them explode when they erupt. Basaltic lava, or basalt, is thin and runny and tends to flow easily, or when it is frothy, to pile up in small pellets as cinder cones.

Textures of igneous rock.

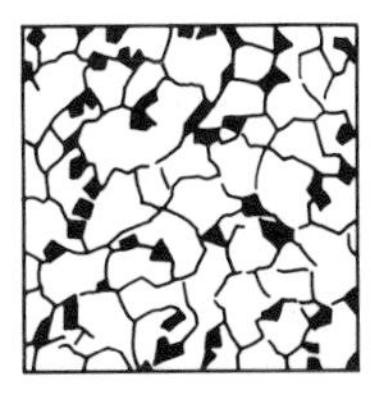

In granite, crystals are all about the same size.

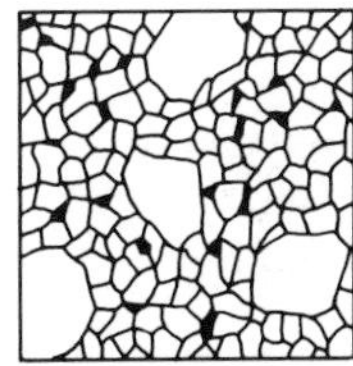

Porphyry contains large crystals scattered in a finer matrix.

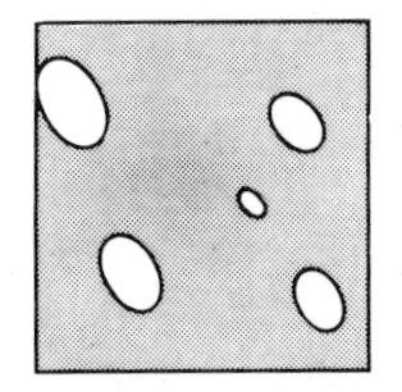

Crystals in volcanic rocks are often invisible without magnification. Gas bubble holes are common.

Prominent layers are typical of sedimentary rocks. Here, weak shale layers weather into slopes; stronger limestone and sandstone layers form small ledges.

If molten magma cools and crystallizes below the surface, it becomes intrusive igneous rock. Individual mineral crystals in intrusive igneous rock grow larger than do those of volcanic rock, so intrusive igneous rock is quite coarse-grained, giving us our word for it: granite.

Sedimentary rocks originate as sediment, the broken-apart remains of older rocks or of animal shells. Sediments can be carried by water, wind, or ice, and are normally deposited in horizontal layers or strata — which makes them quite easy to recognize. Some volcanic rocks, particularly volcanic ash and

Dune sandstone, common in Utah, can be recognized by its long, sloping cross-bedding as well as by its tiny, even-sized, rounded grains.

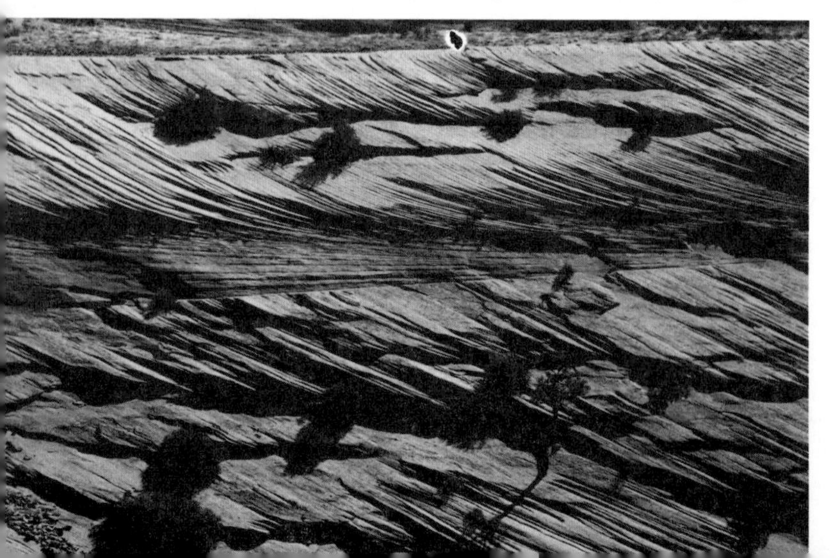

thin basalt lava flows, may be layered, too. Most sedimentary rocks were deposited in lakes, oceans, and river floodplains, though some, such as the coarse conglomerate of alluvial fans or sandstone formed of ancient dunes, accumulated on dry land. Individual layers of sedimentary rock are commonly called beds.

Except for limestone, dolomite, coal, and salt, sedimentary rocks are classified by particle size: Conglomerate contains coarse particles, pebbles and cobbles and even boulders, usually mixed with sand. Sandstone, siltstone, mudstone, and claystone are composed of successively smaller particles. The term shale is used for siltstone or mudstone that breaks easily into flat slabs. Limestone forms from accumulations of calcium carbonate, and is often derived from animal and plant shells and skeletons.

Textures of sedimentary rocks

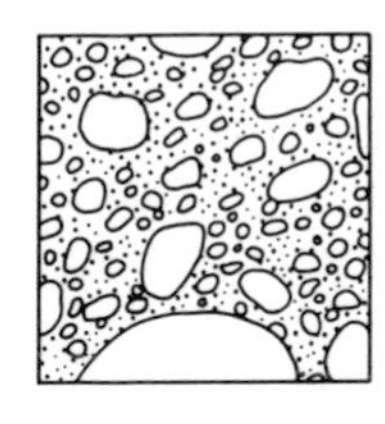

Conglomerate, formed from gravel, contains both large and small rock fragments rounded by bouncing and grinding against one another in running water. The matrix is usually sand.

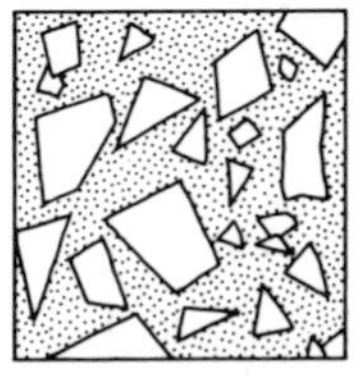

In breccia, rock fragments are angular. Matrix may be clay, finely crystalline rock, or minerals which grew later around the fragments, cementing them together.

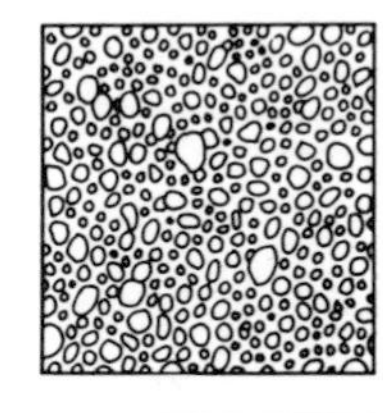

Sand is made up of particles 1/32 to 1/4 inch in diameter. Grains are usually rounded.

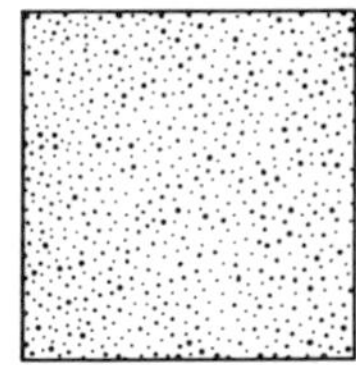

Siltstone particles are less than 1/32 inch in diameter.

When sedimentary rocks are tilted, as many in Utah are, we speak of their dip, the angle between horizontal and their maximum downward slope.

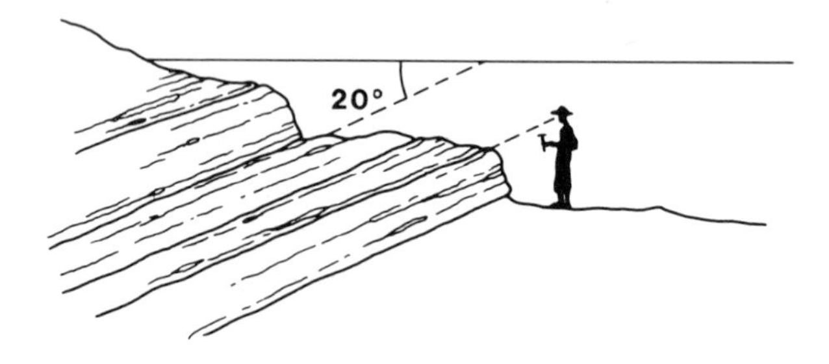

Dip is the angle between horizontal and the downward slope of bedding or stratification.

Metamorphic rocks have been changed from what they were originally. Both heat and pressure are involved in their alteration. Sandstone becomes quartzite, limestone becomes marble, while siltstone and mudstone become slate. With greater extremes of temperature and pressure, siltstone, mudstone, and silicic volcanic rocks completely recrystallize into schist and gneiss. Schist, with lots of mica giving it a tendency to split along parallel planes, generally develops from shale, mudstone, or siltstone. Gneiss, coarse and grainy and looking like striped granite, commonly develops from sandstone or light-colored volcanic rocks or granite. Greenstone, not very common in Utah, develops from basalt lava flows.

All rocks are made of minerals. Different minerals have definite physical and chemical properties by which they can be recognized. This book does not emphasize minerals; if you are interested in them you probably already have one or more of the many guidebooks to minerals.

Minerals are identified in the field by their characteristic colors, hardness, and ways of crystallizing — in six-sided rods or small cubes or nondescript glassy masses, for instance. Parts of Utah yield particularly beautiful minerals, such as azurite and malachite, two ores of copper. These and other minerals are exhibited in museums around the state. Many are available

COMMON ROCKS OF UTAH

		ROCK	DESCRIPTION
SEDIMENTARY		Conglomerate	Sand and pebbles deposited as gravel, then cemented together. May include cobbles and boulders
		Sandstone	Grains of sand cemented together
		Siltstone	Grains of silt cemented together
		Claystone	Clay-sized grains cemented together
		Mudstone	Grains of clay and silt cemented together
		Shale	Flat-slabbed siltstone or mudstone
		Limestone	Rock more than 50% calcite (calcium carbonate)
IGNEOUS	EXTRUSIVE	Basalt	Very fine-grained gray or black volcanic rock, either flows or cinders
		Rhyolite	Light-colored silicic volcanic rock, in short lava flows, tuff (volcanic ash), or breccia
		Dacite	Dark-colored volcanic rock present in some stratovolcanoes
	INTRUSIVE	Granite	Light-colored, coarse-grained rock with visible crystals of quartz, feldspar, and black mica or hornblende
		Monzonite and Diorite	Medium-grained rocks made predominantly of feldspar, with black mica or hornblende and very little quartz
		Porphyry	Any of the above with large crystals scattered throughout a finer groundmass
METAMORPHIC		Marble	Recrystallized limestone
		Quartzite	Sandstone cemented with silica so that it breaks through grains
		Gneiss	Streaky or banded crystalline rock formed from granite or sandstone
		Slate	Hard, fine-grained, slabby rock formed from siltstone, claystone, or mudstone; breaks along planes other than bedding planes
		Schist	Streaky medium-grained rock that tends to split along parallel planes of mica, formed from shale or mudstone

in rock and mineral shops. Utah's most common rock-forming minerals are described below.

Quartz is a glassy or milky white mineral so hard that it scratches glass but can't be scratched with a knife. (Diamonds are not the only minerals that will scratch glass.) Quartz occurs as glassy grains in granite and in sandstone derived from granite. Some varieties, such as rose quartz and amethyst, are tinted with minute amounts of other minerals.

The feldspar family includes translucent pinkish, grayish, or whitish minerals that can, if you press hard, be scratched with a knife. These normally break along flat cleavage faces that reflect sunlight. Feldspar crystals abound in granite and other intrusive rocks, as well as in metamorphic rocks. Because feldspar readily breaks down into clay it is not usually found in sandstone except where the sandstone is very near its source.

The mica family includes a number of shiny black or silvery minerals that separate into shiny, paper-thin flakes. Mica can be scratched easily with a knife, even with a fingernail. Black or brown mica or biotite, and white mica or muscovite are common varieties. They occur in granite and other intrusive rocks as well as in metamorphic rocks such as schist and gneiss.

Calcite is a white or light gray mineral, the chief component of limestone. It can't be scratched with a fingernail, but can be scratched with a knife. When dilute acid is dropped on calcite, it fizzes.

Hematite is red iron oxide, and is easily recognized by its dark rust-red color. In tiny amounts it lends pinks and reds to many of Utah's most scenic areas.

Limonite is a rusty yellow iron oxide that in small amounts tints rocks yellow or tan.

Gypsum is translucent or white or gray, and so soft that you can easily carve it with a knife. Many of eastern and southern Utah's red rocks are veined with white selenite, a variety of gypsum.

Salt crystallizes when sea water or salty lakes dry up. It exists in many parts of Utah, not only around Great Salt Lake but also deep underground in much older rocks throughout eastern Utah.

JOINTS, FAULTS, AND FOLDS

Sets of parallel cracks, called joints, are common in most rock; many rocks display several intersecting sets of parallel joints. Some joints form because fine sediments such as clay shrink as they dry; others form because molten rock, or magma, shrinks as it crystallizes. Other joints develop because the stresses involved in movements of the Earth's crust simply break the rocks. Joints form pathways for water and air, and therefore facilitate weathering of rocks into soil.

Faults are breaks or fractures along which rocks on one side have moved relative to those on the other. Movement may range from a few feet to many miles.

Faults are classified by the angle of the fracture surface, the fault plane. The accompanying diagrams show several kinds of faults. Normal faults and detachment faults generally form where tension pulls the Earth's crust apart. In contrast, reverse and thrust faults form by compression. Strike-slip faults move horizontally, one side past the other. Faults usually show up best in sedimentary rocks, where clearly defined layers are offset, and in regions of varied rock types, where one ends abruptly, butting into another. They are harder to pinpoint in large masses of igneous or metamorphic rocks, where the rocks on opposite sides may look the same despite large displacement. Most faults are found by careful mapping of rock types and ages, the first step in understanding the geology of any area.

Large faults with considerable displacement on each side are rarely simple breaks. They commonly come as fault zones, in which numerous small, more or less parallel faults slice through the rock. Or the rock in fault zones may be completely shattered, ground into a flourlike mass by movement of rock grinding against rock.

Utah lays claim to many thousands of faults, large and small. In the central mountains of the state, one particularly prominent thrust fault carried a relatively thin sheet of sedimentary rocks eastward over younger rocks. This fault is usually referred to as the Sevier thrust belt; you will cross it on some of the highways discussed in this book.

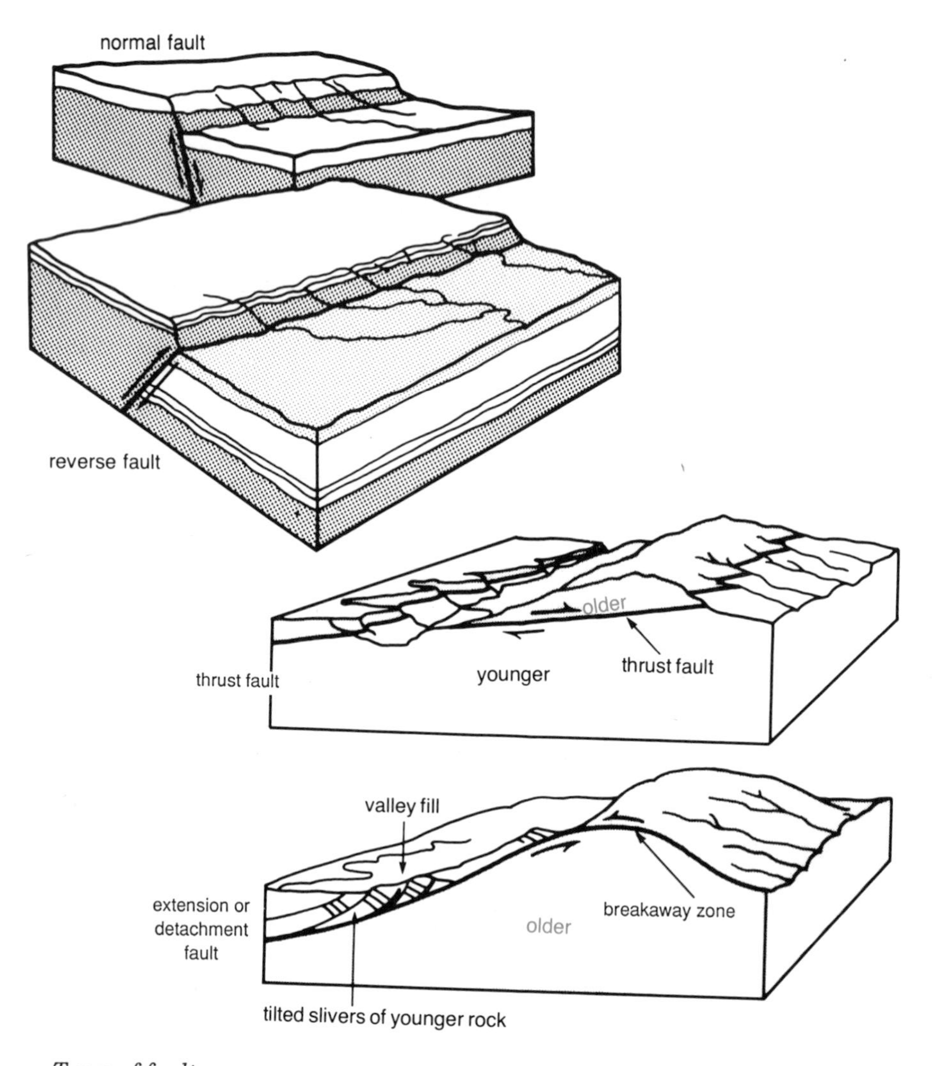

Types of faults. —Reprinted by permission, from Pages of Stone: Geology of Western National Parks and Monuments, 3, THE DESERT SOUTHWEST, by Halka Chronic (The Mountaineers, Seattle).

Major normal faults edge most of the mountain ranges in the western part of Utah. Growing evidence indicates that these steeply inclined faults level out as they get deeper, and join nearly horizontal or gently undulating detachment faults.

Utah's earthquakes show that some of the state's many faults are still active.

Folds are just that: folds or bends in the rock. It seems strange that rocks can bend, but given plenty of time and

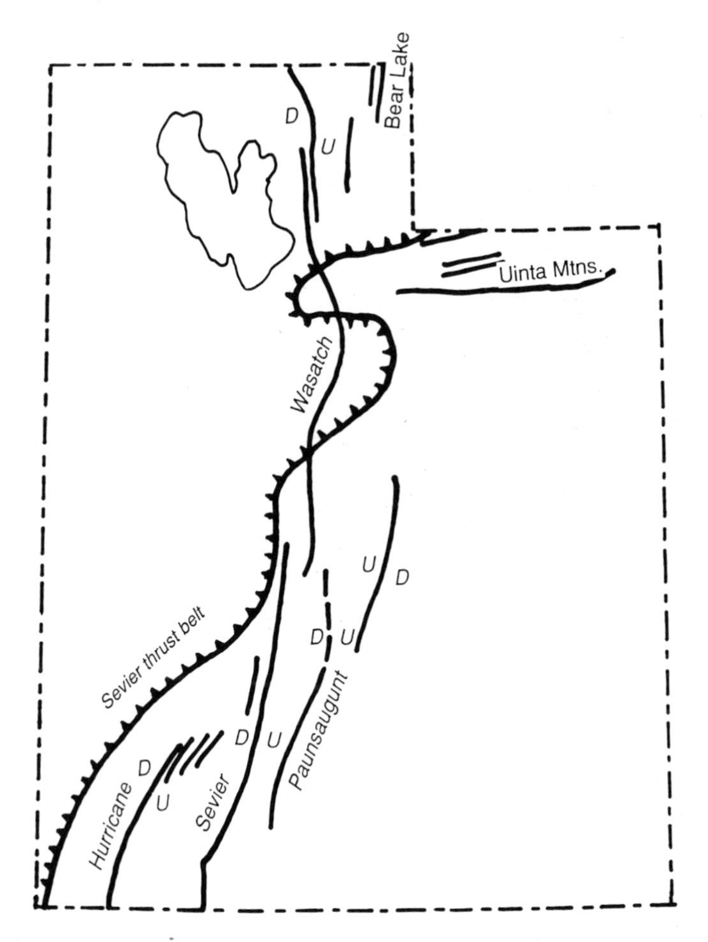

The largest of Utah's many faults run roughly north-south through her central mountains. Thousands of other faults have been mapped here and elsewhere in the state. D=down, U=up. On thrust faults, teeth are on the upper or overriding plate.

stress, they can. Some folds are very slight — sort of wide-open bends that are nothing more than a gentle warping of the rock layers. Others are like accordion pleats — tight to the point where the two sides of any one fold are parallel. Most folds fall between these extremes.

Folds are classified as anticlines or upward folds, synclines or downward folds, and monoclines, one-way folds in which rock layers on one side are raised higher than corresponding layers on the other side. Like faults, folds are most easily detected in sedimentary or layered volcanic rocks. Most are

This large monocline occurs along the south edge of the Uinta Mountains east of Vernal, Utah, where Pennsylvanian sandstone bends down off the mountain and into the Uinta Basin.

products of horizontal compression or squeezing of part of the Earth's crust. Monoclines may form as near-surface rocks drape over deeply buried normal or reverse faults.

BACK TO PLATE TECTONICS

Utah lies far inland, not particularly close to scenes of violence along the margins of the North American plate. Nevertheless, it displays some of the features normally associated with plate margins: deep sedimentary basins, mountains, volcanoes. As we'll see, Utah's western ranges are thought to have formed as the continental crust stretched.

Utah seems to lie on a sort of hingeline which in Precambrian and Paleozoic time ran almost parallel to the position of the Earth's equator, along what was then the continental margin. Today, thanks to counterclockwise rotation of the continent, that old hingeline runs about north-south, almost parallel to Interstate 15 from Salt Lake City to St. George. East of the hingeline (south of it in Paleozoic time), on the stable shelf of the continent, only thin sediments were deposited — mostly marine in late Precambrian and Paleozoic time, mostly continental in Mesozoic and Cenozoic time. West of the hingeline the continent sagged, forming a deep basin, and sediments deposited in it were much thicker — measured in thousands of feet — than those east of the hingeline. Throughout late Precambrian and Paleozoic time, and well into Mesozoic time, the edge of the continent was much closer to Utah than it is now.

Late in Paleozoic time, two smaller subsiding basins developed within Utah: the Oquirrh Basin of northwestern Utah, which filled with fine-grained sediments now known as the Oquirrh group; and the Paradox Basin of southeastern Utah, which filled with rock debris from an adjacent highland, as well as with thick layers of salt, gypsum, and potash deposited as sea water evaporated.

With formation of new crust along the mid-Atlantic ridge, the North American plate has moved westward (arrow) across the East Pacific plate, now mostly hidden under western North America. Similarly, the South American plate is overriding the Nazca plate. Mid-ocean ridges are offset by numerous transform faults.

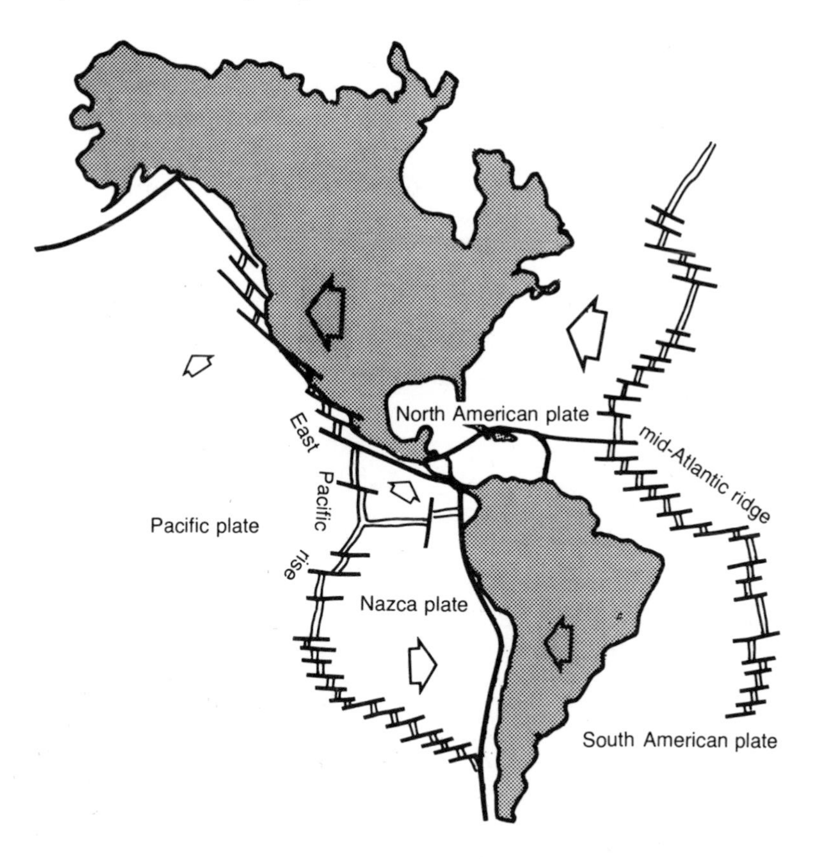

During the Mesozoic era, continental (non-marine) deposits accumulated in eastern Utah, some of them on a sand-swept desert about the same size, and then at about the same latitude, as today's Sahara. Other sediments were deposited in marshes or on river floodplains or in landlocked embayments of a shallow sea.

Late in the Mesozoic era, the Sevier Mountains rose in western Utah, to shed gravel, sand, and silt east into the central part of the state. The buckling and thrust faulting that created these mountains made Utah narrower by some 40 to 60 miles. Apparently powered by compression from the west, the mountain building created Utah's Sevier thrust belt, with giant thrust faults that brought the Oquirrh group and older Paleozoic sedimentary rocks some 40-60 miles eastward over younger Mesozoic rocks. Many geologists think that this thrusting was closely related to the buildup of mountains in California, which in turn related to the way the East Pacific plate sank downward as the North American plate collided with it and rode southwestward over it.

During this time, several islands or minicontinents collided with North America's west coast as the continent rode over the sea floor. Some of these additions to the continent, born far out in the proto-Pacific, can be recognized today in the ranges of California, Nevada, and Idaho.

As the Sevier Mountains rose, a sea once more spread across eastern Utah — this time advancing from the east. In it were deposited marine or near-shore siltstone, mudstone, and sandstone; layers of coal formed in lagoons and swamps along a low, gently sloping coast.

As the Mesozoic era drew to a close, more mountain building changed the shape of the land east and north of Utah. In Utah and Wyoming the Rocky Mountains began to rise, in an episode of mountain building now called the Laramide Orogeny. Among the ranges of the Rockies were Utah's Uinta Mountains — like many other Rocky Mountain ranges a single arched anticline broken along both edges. This mountain-building, too, must be related to plate tectonics, though there is still uncertainty as to how it ties in with the history of the rest of the continent. The North American plate had by this time completely broken its ties with Europe and Africa; several studies show that the timing of the Laramide Orogeny coincides with a sudden change in direction of plate movement and an increase in the speed with which the North American plate traveled southwestward.

One of the results of the Laramide Orogeny was the development of broad lake basins between the new Rocky Mountain ranges and rising mountains in central Utah. Soft siltstones and silty limestones deposited in a succession of lakes during the early part of the Cenozoic era now appear in cliffs and ridges in eastern Utah. Some are scenery-formers; others contain rich deposits of oil shale or beautifully preserved fossil fish.

About 40 million years ago, volcanoes began to erupt within what would become Utah. Thick, sticky silicic magma generated from remelted light-colored continental rock built up towering volcanoes, and immense explosions showered volcanic ash across the land. Around 18 million years ago, stretching, thinning, and breaking of the Earth's crust began to create the Basin and Range mountains of western Utah, Arizona, Nevada, and California.

Studies of earthquake waves show that the crust is thinner in Basin and Range areas than it is elsewhere. Utah's share of the Earth's crust at this time was stretched by at least 50 miles, roughly making up for the 40- to 60-mile shortening of the Sevier Orogeny in Cretaceous time. Taken as a whole, the Great Basin has about doubled in width.

Let's look for a moment at the Basin and Range region and the strange, almost unbelievable geologic events involved in creating it. In Utah, most of the ranges are steeply faulted along their western sides. Geologists soon likened this region to a pile of dominoes all tilted in the same direction — a fairly simple picture. Increasingly, however, new evidence shows that this region is far more complex than a pile of dominoes. Studies of surface features and of manmade seismic waves reflected from deep rock layers show that the steep normal faults that edge the mountains flatten with depth. They seem to merge with other faults to become gently undulating and nearly horizontal detachment faults along which the upper and lower parts of the crust pulled apart. In several so-called metamorphic core complex mountain ranges, including the Grouse Creek, Raft River, and Mineral Mountains, the detachment faults appear at the surface, and can be mapped and studied in some detail. These ranges appear to have bobbed up — on a huge, slow-moving scale — as overlying rocks, above the detachment faults, slid westward.

What caused the great stretching that created the detachment faults? Many geologists believe it resulted from gravity-induced spreading of the overthickened crust of Mesozoic time — thickened by sideways compression and development of the Rocky Mountains. Other geologists propose that the ranges and basins are due to faulting associated with gradual upwelling in the depths of the mantle. Yet others think they may have resulted from the way in which the North American plate is overriding the East Pacific plate, and from the pull exerted by northward drift of the Pacific plate — the same movement that created the notorious San Andreas fault in California. All these processes, and perhaps others, may have helped create the basins and ranges.

At about the same time as detachment faulting, and perhaps partly responsible for it, Utah and adjacent states bowed upward in a broad arch. Utah rose as much as 5000 feet. Their steepened gradients made the streams erode more deeply, cutting down into the lake sediments of eastern Utah, removing the arched sedimentary rocks from mountain summits, carving the myriad canyons of the Plateau Country, filling intermountain valleys or carrying rock debris away down the newborn Colorado River. Erosion whittled Utah's high central backbone and sharpened desert ranges to the west, where heavily loaded streams unceremoniously dumped their debris in desert basins between the ranges.

Only in the intermountain basins was deposition still taking place. In Pleistocene time, the heavy snows and rains that came with the ice ages augmented erosion. Abundant rains and melting mountain glaciers filled the deepest basin, right next to the Wasatch Front, creating Lake Bonneville, the freshwater ancestor of today's Great Salt Lake. As a final touch, Lake Bonneville etched its signature in horizontal shorelines cut into surrounding mountain slopes.

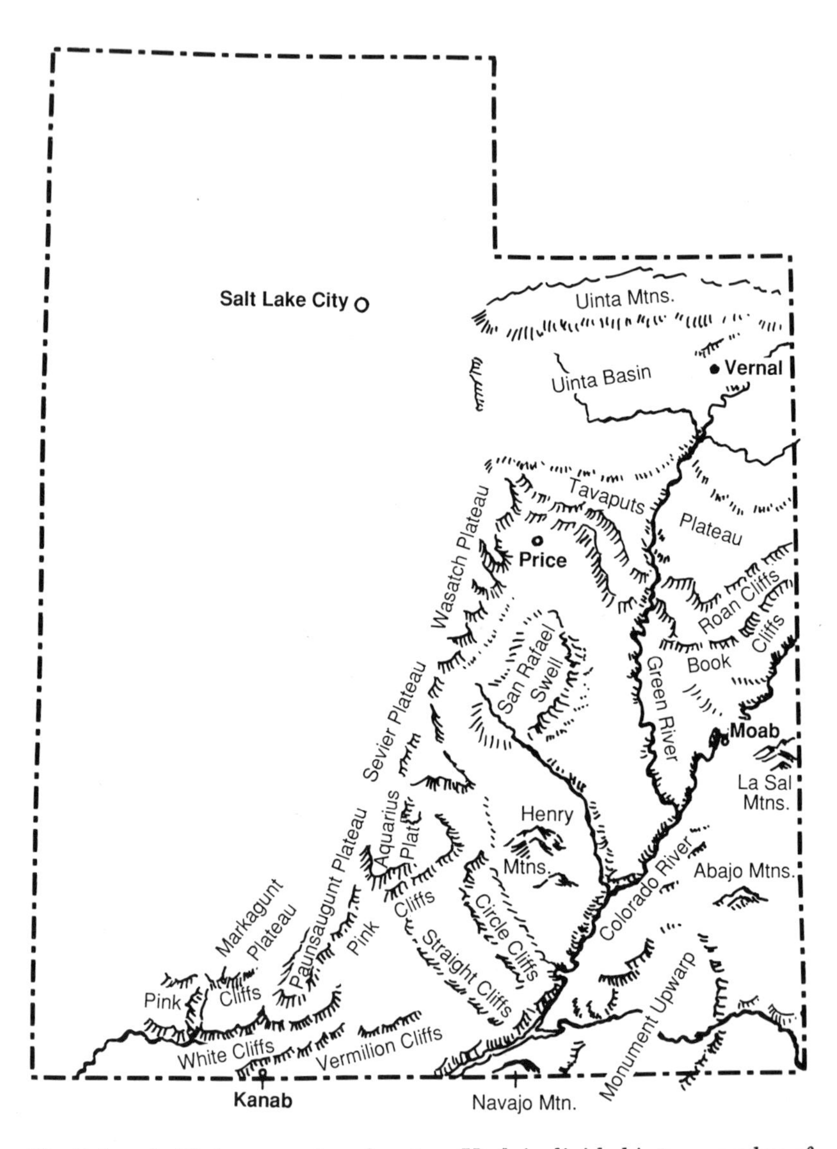

The Colorado Plateau country of eastern Utah is divided into a number of plateaus, basins, and upwarps, natural divisions outlined by lines of cliffs or hogbacks of tilted rocks, and by abrupt changes in elevation. It's not hard to see why the southern part of this area has been termed a Grand Staircase.

I

High, Wide, and Lonesome

The Plateau Country

Most of southern and eastern Utah fits into a geographic and geologic province known as the Colorado Plateau. Here, sedimentary rocks, flat-lying but offset vertically by faults and folds, make up most of the surface. Thanks to an arid climate and deep downcutting by the Colorado River and its many tributaries, the rocks are well exposed, revealing themselves in characteristic layer-cake patterns of sedimentary rocks that have not been tilted, distorted, or altered by mountain building. Since they are so well exposed, and since many of them are easily recognized, it seems especially appropriate in this chapter to call them by name.

Formations — recognizable, mappable rock units — and groups of formations are named for geographic sites where they are well and typically exposed. Thus the Bluff sandstone is named for the town of Bluff, the Green River formation for the Green River, the Paradox formation for Paradox Valley, all in the Plateau country of eastern Utah. Some names used in Utah come from other states — the Dakota sandstone and Mesaverde group and Kayenta formation, for instance — and help to indicate the wide distribution of certain rock units.

Named not for the state of Colorado but for the river that courses through it, the Colorado Plateau is more or less oval in outline. The Uinta Mountains lie along its northern edge, the

Rockies to the east, and the Wasatch Mountains to the west. Almost all of it drains through the Colorado River and its tributaries. Extending south into Arizona and New Mexico and east into Colorado, the entire region seems to have been by-passed by mountain building, at least by the kind of mountain building that crumples and melts and accordion-pleats parts of the Earth's crust. Faults and simple folds divide it into a number of smaller plateaus, as well as into the two broad anticlines of the San Rafael Swell and the Monument Upwarp. Northward it includes the Uinta Basin — a sagging, downwarped section of the crust. The westernmost of the smaller plateaus — the Paunsaugunt, Aquarius, Markagunt, and Wasatch plateaus — are several thousand feet higher than those farther east; they are here considered part of Utah's High Country, discussed in Chapter II. The oldest sedimentary rocks of the region are Paleozoic, exposed where they dome upward in the Monument Upwarp and the San Rafael Swell, along the southern edge of the Uinta Mountains, and in smaller anticlines near Utah's eastern border. Mesozoic sedimentary rocks, Triassic, Jurassic, and Cretaceous, cover much of the remaining area, particularly in the southeast quarter of Utah. Cenozoic rocks appear on the Roan Plateau north of the towns of Green River and Price, in the Uinta Basin, and around the southern fringe of Utah's High Plateaus.

The layered rocks of the southern part of this region are said to form a Grand Staircase, a succession of cliffs rising northward from the Arizona border. The successive risers are the Vermilion Cliffs, the White Cliffs, the Gray Cliffs, the Pink Cliffs, composed respectively of resistant Triassic, Jurassic, Cretaceous, and Tertiary sedimentary rock layers. In places the uppermost tread is a final resistant lava cap. The steplike plateau surfaces range in elevation from 4500 to 10,000 feet above sea level.

One of the nice things about the whole Plateau region is the color of the rocks: pink and red, yellow and green, purple and white. They make America's most colorful landscape. The colors come from minor constituents of the rock: reds and pinks and yellows created by tiny particles of iron oxides, greens and blues of unoxidized iron minerals, lavender of manganese. In addition there are the contrasting gray of igneous rocks, basalt

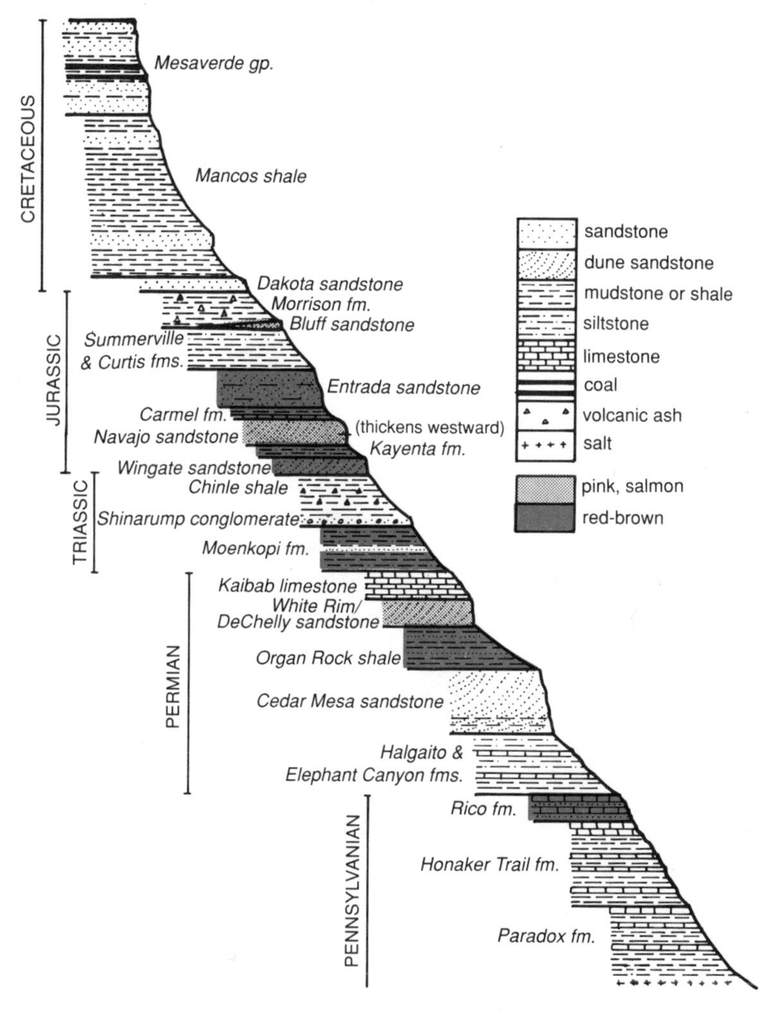

Generalized stratigraphic diagram of Paleozoic and Mesozoic sedimentary rock units exposed in Utah's Plateau country.

and granite, and dark browns and purples of desert varnish, a dark shiny surface that forms on rocks in desert regions. Lichens color some rock surfaces with green and orange blotches, or streak cliffs with long ribbons of brown and black.

Here in the rainshadow of the Utah high country, precipitation is scanty. Soils are thin and vegetation sparse. Many landforms result from differential weathering and erosion of hard and soft rock layers, with hard layers (usually sandstone and limestone) standing up as cliffs and ledges, soft layers (usually mudstone and shale) forming slopes and benches.

Wind is an important agent of erosion, hammering exposed rock with tiny tools of sand, smoothing rock surfaces, chiseling rounded holes, shaping rock arches and balanced rocks, and in summer lifting the debris high in swirling dust devils. Together with rain, wind carries away sand and silt, leaving behind only a desert pavement of closely spaced pebbles, an armor that retards further erosion.

We've mentioned that the individual small plateaus that make up the Colorado Plateau are edged with cliffs and hogbacks. In some cases these mark fault zones, and differences in elevation are due to fault movement. In other cases the plateaus are delineated with monoclines, the simplest of folds, with one side raised higher than the other. Many such monoclines are thought to be a "draping" of sedimentary rocks over deep faults. Elsewhere, particularly in the Grand Staircase, the differences in plateau levels are the work of erosion, the stripping back of rock layers.

Much of the shaping of the Plateau country has occurred in the last 10 million years, during and after regional uplift of Utah and its neighboring states. Uplift added new strength to streams that earlier wound sluggishly across a broad, gently sloping terrain surfaced with soft Tertiary sediments. Most of the Tertiary sediments washed away, leaving the streams and rivers entrenched in harder Mesozoic and Paleozoic rocks. There, cutting constantly downward, they commonly retained their winding courses, incising their old meanders into the Mesozoic and Paleozoic rocks, giving us much of the scenery we see in the Plateau country today.

Here and there in this area there are signs of igneous activity: mountains that are clustered intrusions; stark volcanic necks, the conduits of former volcanoes; dikes of hard igneous rock that jut as ridges above the ground surface; lava-capped plateaus. But this is mostly a land of sedimentary rocks, Mesozoic sedimentary rocks with touches here and there of Paleozoic strata exposed in eroded anticlines. Quaternary sediments, not yet consolidated into rock, include alluvial fans, stream floodplain deposits, a few glacial moraines on the high peaks of the La Sal Mountains, and wind-deposited sand and silt on some plateau surfaces.

The Plateau country is a geologist's heaven, not just with its clearly exposed rock layers and structures but also with its bountiful quantities of oil, natural gas, and uranium. Of gold and silver there is little, though copper is mined in Lisbon Valley near the Colorado border. Vast reserves of petroleum are tied up in oil shale and tar sand, awaiting the day when the price of oil rises high enough to make their processing profitable.

Just how the Colorado Plateau fits into the plate tectonics pattern is hard to say. As described above, the plateau consists of many individual segments — smaller plateaus and basins that rose and fell relative to each other. The faults and monoclines, broadly speaking, trend north-south. Taken as a whole, the Plateau region seems to have rotated clockwise relative to other parts of the continent, a rotation that may be linked to northward movement of the Pacific plate and the drag that movement exerts on the west edge of the North American plate. But why the Plateau retains its integrity, like a raft in a stormy sea of mountain-building, has not been satisfactorily explained.

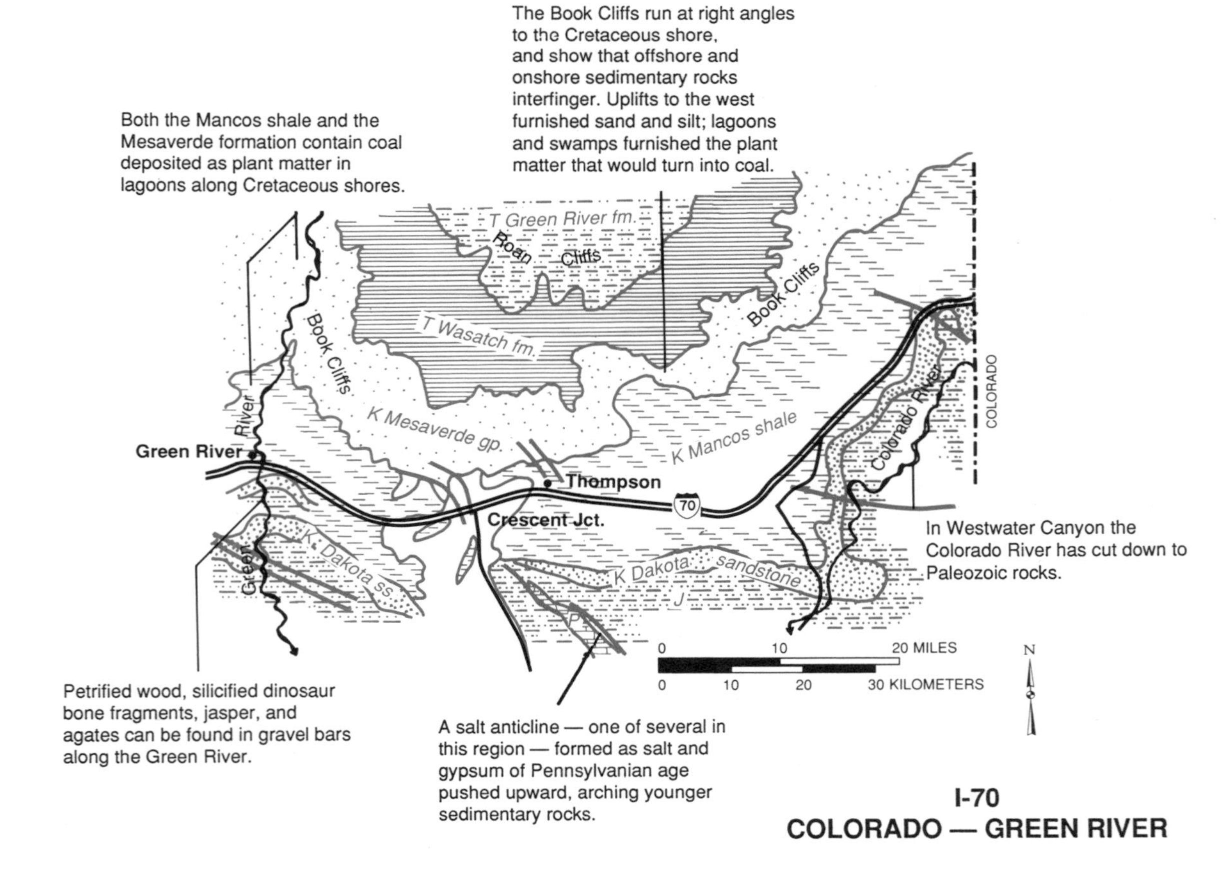

I-70
COLORADO — GREEN RIVER

The Dakota sandstone includes cross-bedded sandstone deposited in tide-scoured channels along the shore of an advancing Cretaceous sea. Water-formed cross-bedding is smaller in scale than dune cross-bedding.

Interstate 70
Colorado — Green River
90 miles/145 km.

At the Colorado-Utah state line, the highway, curving around the north end of the Uncompahgre Uplift, rides on a dip slope of Dakota sandstone, the oldest Cretaceous formation throughout the Plateau country. The Dakota sandstone, as well as other sedimentary rocks in this area, dips northward off the Uncompahgre Uplift. Both the Dakota sandstone and the Morrison formation underlying it can be seen at the view area at milepost 226.

The highway soon leaves the Dakota hogback, which swings away southward to complete its circling of the Uncompahgre Uplift. From here to Green River the highway stays on the Mancos shale, a soft, fine gray shale deposited as mud and silt in the Cretaceous sea. To the north are the Book Cliffs, composed of sandstone, shale, and coal of the Mesaverde group, which is also Cretaceous. Hidden behind them, above a broad bench of Tertiary Wasatch formation lake deposits, are the Roan Cliffs, consisting of the Green River shale, deposited in a later Tertiary lake.

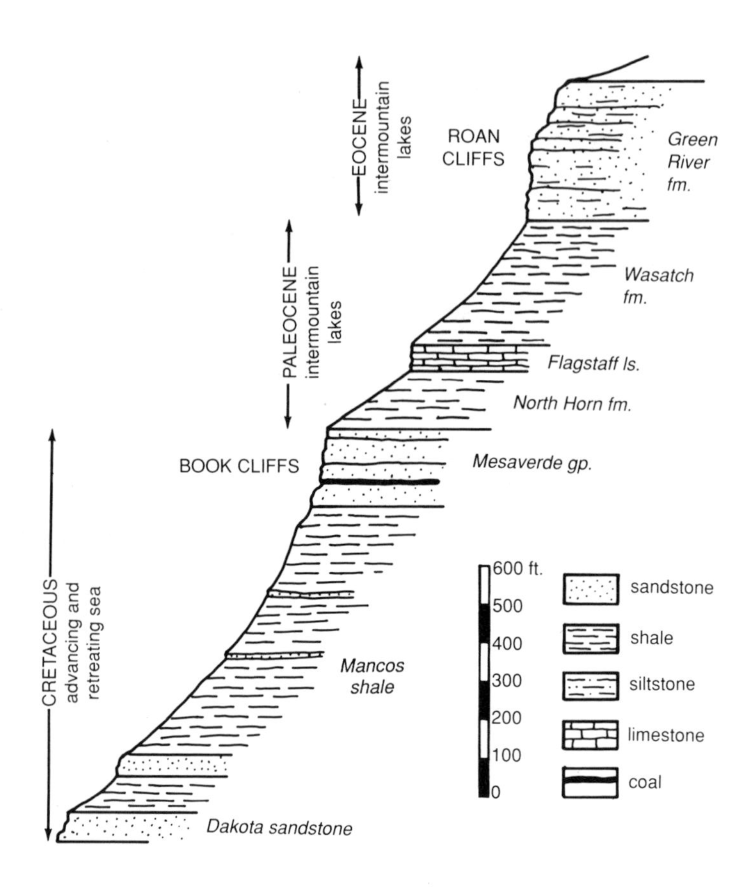

Stratigraphic diagram of Book Cliffs and Roan Cliffs north of Interstate 70.

Some of the highway cuts, particularly those between mileposts 226 and 224 near the junction to Westwater, reveal layers of coal.

There are good views to the south of the arched tableland of the Uncompahgre Uplift, a high fault-edged plateau that straddles the Colorado-Utah boundary line. The Colorado River circles north of this uplift, not quite managing to avoid its northern end. In Westwater Canyon the river has carved all the way down into Precambrian rocks that form the core of the uplift. Jurassic rocks of the Uncompahgre Uplift yielded one of the largest dinosaurs known, a giant that stood as high as a seven-story building!

Let's take another look at the Book Cliffs. In the flat-lying rock layers of the Plateau country, cliffs erode from the side rather than from the top. As weak rock layers are washed away, harder ones are undermined, and eventually break away in slabs large and small.

Over thousands and even millions of years, the cliffs retreat — in this case northward. The cliffs were once where the highway is now.

The route crosses many washes that drain Book Cliffs, all of them dry most of the time. When they do flow, after sudden storms, they carry away some of the rock debris worn from the cliffs. Whitish stains on the soil surface show that there is a considerable amount of alkali and salt in the soil here; evaporation of soil moisture leaves the white substances on the surface. Steep-sided gullies along some of these washes are fairly recent additions to the landscape, having formed in the last 100 years. They seem to result from grazing by sheep and cattle, which removes protective plant cover and exposes soil to erosion.

Southward there are good views of the La Sal Mountains, a cluster of Tertiary intrusions about 35 million years old. Permian, Triassic, Jurassic, and Cretaceous sedimentary rocks pushed upward by the rising masses of magma have now eroded away, but their edges remain in irregular rings around the intrusions. You can see some of them in profile on either side of the mountains. And from milepost 215 you can see the gorge of the Colorado River, almost a Grand Canyon in its own right, as well as the slot west of the La Sals through which the river flows. Some remarkable spires can be seen from here, close to the La Sals — tall, slender pinnacles of rock.

Isolated oil wells between the highway and the Book Cliffs obtain oil from a Pennsylvanian formation, the Weber sandstone. Some wells produce helium, and are part of the U.S. Helium Reserve.

Lumpy pink rocks on the skyline west of milepost 194 are in Arches National Park. They are the Jurassic sandstone fins in which the famous arches formed. As the highway converges with the Book Cliffs, we get better and better views of them.

The rest area west of milepost 188 provides a good place to stop and look at both the Book Cliffs and the Roan Cliffs. The Tertiary rocks of the Roan Cliffs, the Green River formation, were deposited in large lakes between rising mountain ranges: the Rocky Mountains in Colorado, Wyoming, and Utah. The Wasatch formation, which forms the bench between the Book Cliffs and the Roan Cliffs, is also a lake deposit.

The viewpoint is surrounded with hilly badlands of tan and gray Mancos shale, the deposits of a sea that covered the western interior of North America in Cretaceous time. Some of the badland hills are capped with sandstone of the Mesaverde group, the lowest rock unit of the Book Cliffs. The Mesaverde group contains some lower sandstone layers, then a weak layer of siltstone and coal, and then more

North of the highway, the Book Cliffs, formed in layered rocks of the Mesaverde group, rise above rounded hills of gray Mancos shale.

sandstone layers, all deposited along the shore of the Cretaceous sea as it shallowed and retreated eastward. The sandstones are beach and bar sandstone; the coal accumulated in marshes and swamps where plant material became a large part of the sediments. Dinosaur tracks found in this area show another giant, with a 15-foot stride.

The Green River formation is famous for thick layers of oil shale and for its particularly fine and abundant fossil fish. Something like 1.8 trillion barrels of oil reserve exist in the oil shale, making it the richest such deposit in the world. But as yet there is no profitable method of extracting it. The oil exists in the form of kerogen, a waxy substance, tightly held in extremely small pore spaces in the shale. Its extraction involves heating the shale so that the oil can flow out of it, an expensive process. And of course mining and processing the oil shale must also be done in a way that is environmentally sound. Disposal of a trillion-odd tons of spent shale will be a major problem.

If you look straight south from mileposts 181-180 you will see the Salt Valley anticline, on which the arches of Arches National Park have developed. As its name suggests, this anticline was formed by salt. Thick layers of salt, deposited near the edge of a Pennsylvanian sea, underlie much of eastern Utah and western Colorado. Because of varying pressures in the rocks above it, the salt tended to flow very slowly from areas of greater pressure to areas of lower pressure, here toward some buried ridges of Precambrian rock. As the salt flowed toward these ridges, it pushed up overlying sedimentary rock, forming long, narrow anticlines like the one you see here. Eventually, as the salt was dissolved and carried away by groundwater, the crests of the anticlines collapsed.

Mines near Thompson, north of the highway, produced uranium from the Morrison formation, the Jurassic rock layer just below the Dakota sandstone. Dinosaur tracks were found in one of the uranium mines here.

As you have probably noted, areas surfaced with Mancos shale support few plants. The problem is partly the arid climate, five to six inches of rainfall annually, partly the high selenium content of the Mancos shale. The rock contains types of clay that swell and shrink with each wetting, making it hard for plants to establish themselves.

The long, low mountain visible to the west as the highway approaches Green River is the San Rafael Swell, an anticline 100 miles long from north to south and up to 40 miles wide. Its crest, eroded down to Permian limestone, is surrounded by ridges of tilted Triassic and Jurassic sedimentary rocks, some of which form pointed flatirons (yes, that's a geologic term) around its flanks.

In the distance to the southwest are the Henry Mountains, like the La Sals a cluster of small intrusions.

Near Green River, the Book Cliffs swing away to the north, as do the gray and mustard-colored hills of Mancos shale. Their yellowish color comes from rapid weathering, with oxidation of iron minerals. Some of the hills are beveled and wear a topcoat of red-brown gravel swept downward and outward from the Book Cliffs and Roan Cliffs.

Edged with trees, the Green River flows fairly passively here. Upstream, it flows out of Desolation Canyon, named in 1869 by John Wesley Powell, a Civil War veteran and geologist who boldly led the first exploratory float trip down the Green and Colorado rivers. Heading in the Wind River Mountains of Wyoming, the Green River is by far the largest of the Colorado's tributaries, often carrying more water than the Colorado River.

Petrified wood, silicified dinosaur bone fragments, jasper, and agates can be found in gravel bars along the Green River. There is uranium in this country, too, in river-deposited sandstone of the Jurassic Morrison formation. Long before the uranium boom, in the early 1900s, this area also saw some radium mining.

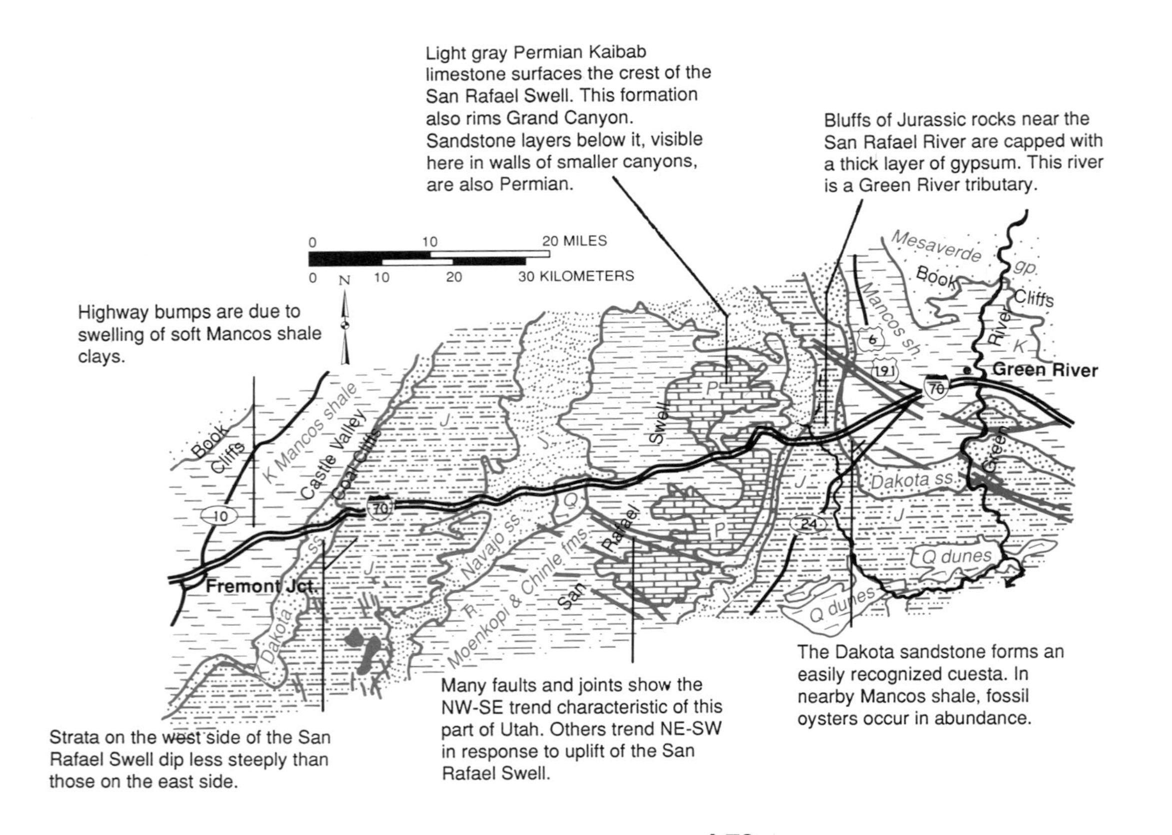

I-70
GREEN RIVER — FREMONT JUNCTION

Interstate 70
Green River—Fremont Junction
73 miles/117 km.

There is no better place to look at Triassic, Jurassic, and Cretaceous rocks than the San Rafael Swell, the long, low mountain west of Green River. Bowed up in Eocene time into a blisterlike anticline 100 miles long from north to south and about 40 miles wide, the sedimentary rock layers that form it are beautifully exposed. As the highway crosses the anticline it travels through successively older and older rock layers to the summit of the swell, then through successively younger layers to the other side. All the formations are clearly exposed; most of them show characteristic features and ways of weathering that can be recognized in many other parts of the Plateau country.

At the west edge of Green River, I-70 rides on Mancos shale, the soft Cretaceous unit that floors the valley of the Green River. The formation contains dark gray marine mudstone, coal, and cross-bedded near-shore sandstone. In places it also contains fossil mollusks and ammonites. A prominent bed of oyster shells near its base, at about milepost 150, forms a line of low, rounded knolls. Fossil oysters, which indicate a near-shore environment, are in places so abundant at the base of the Mancos shale that they are quarried for road material.

Near the interchange for Utah 24, watch for a sloping ridge, or cuesta, of Dakota sandstone, the oldest Cretaceous unit here, deposited along the shore of the advancing Cretaceous sea. West of the Dakota ridge, pastel-colored shales of the Jurassic Morrison formation make up low, hilly terrain. This formation was deposited in a river delta and swampy lake environment, where decay of abundant plant material prevented oxidation of iron minerals—hence its soft green and purple tones. Sinuous Jurassic stream channels containing light gray sandstone now cap many ridges in this vicinity.

Near the bridge across the San Rafael River, castlelike gray and tan siltstone cliffs of the Jurassic Summerville formation are capped with a massive, thick bed of gypsum. The siltstone and gypsum are thought to have been deposited on tidal flats along the shores of a desert-bordered sea that stretched from here to northern Canada, in an environment not unlike that of the Red Sea or Persian Gulf today. The pinkish sandstone and red-brown mudstone of this formation contain ripplemarks and cross-bedded sandstone lenses typical of those forming today along tide-swept channels. The gypsum layer near the top of the formation, as well as thin gypsum veins that penetrate the siltstone, originated as sea water evaporated in isolated embayments. The sea must have left salt deposits, too, but the more soluble salt has long since leached away.

West of the San Rafael River, pale greenish rocks exposed south of the road are the Curtis sandstone, a thin marine sandstone, followed by the next older formation, the Entrada sandstone. Both are Jurassic, but older than the Morrison and Summerville formations. A resistant cliff-former in some parts of Utah, the Entrada sandstone is quite soft here, and erodes into a wide racetrack valley that circles the San Rafael Swell — a racetrack more than 200 miles around!

The eastern flank of the San Rafael Swell looms higher as we approach it, and reveals the formidable sandstone hogback, locally called a "reef," jutting 800-2000 feet high, that edges it. This virtually impregnable wall made the entire anticline relatively inaccessible before I-70 was constructed. Even now, the interstate is the only highway that crosses it.

At the rest area near milepost 144 are exposures of the Carmel formation, the unit that forms the small outermost cockscomb at the base of the San Rafael Swell. Like the Summerville formation, the Carmel formation was deposited in a desert environment, and contains gypsum left behind by evaporating sea water. It, too, is Jurassic.

Along the east side of the San Rafael Swell, the Navajo sandstone rises abruptly from beneath younger rocks. Interstate 70 squeezes between the "teeth" of the San Rafael hogback.

Where the highway penetrates the Navajo sandstone wall, an angular cliff of Jurassic Wingate sandstone towers over ledges and slopes of older Triassic rock. The same sharp peak shows near the highway in the preceding photograph.

West of the rest area the highway heads between the much larger cockscombs, or hogbacks, formed in white, cross-bedded Navajo sandstone, a Jurassic formation that weathers into rounded cliffs and knobs, and the red Jurassic Wingate sandstone, which forms sharp-edged cliffs dark and shiny with desert varnish. Both of these units are eolian (wind-deposited) sandstone, marked by broad, sweeping cross-bedding of ancient sand dunes. The dune sandstones point to the existence of a Jurassic desert of about the same size and latitude as today's Sahara, along the southern shore of a Jurassic sea. Since it was deposited, the continent's drift has carried the desert dunes far north of their original latitude, rotating them at the same time. Separating the two units are thin layers of flat, slabby, stream-deposited siltstone and sandstone, the Kayenta formation.

West of the big roadcuts we cross another racetrack valley, this one in dark brick-red Triassic rocks of the Chinle and Moenkopi formations. Both consist of weak, easily eroded mudstone and siltstone, though the Chinle also includes sandstone, and there is one prominent limestone layer in the Moenkopi formation. Uranium in the Chinle formation is concentrated in organic material such as decaying leaves and wood, mostly in channel sandstone. Watch for its yellow color.

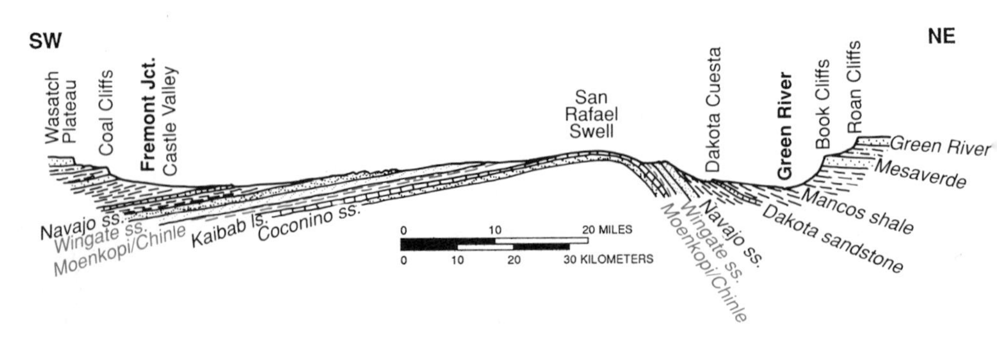

Section along I-70, crossing the San Rafael Swell between Green River and Fremont junction.

As the highway climbs toward the crest of the San Rafael Swell, it emerges from the Triassic redbeds onto light gray or buff Permian limestone, the Kaibab formation,well exposed around the rest area west of milepost 140. From the rest area the Coconino sandstone, the oldest rock exposed on the San Rafael Swell, is visible in canyons north of the highway — more dune sandstone, deposited along Permian shores. Both the Coconino sandstone and the Kaibab formation extend southward into Arizona, where they make up the uppermost part of Grand Canyon's walls.

The crest of the San Rafael Swell, between mileposts 135 and 133, is on essentially flat-lying Kaibab limestone; ledges of the same formation appear in nearby canyons, with white Coconino sandstone below them.

West of the summit, we descend through the same rock units we've seen, but in reverse order, with the strata getting younger westward. Since the rocks dip more gently on the west side of the San Rafael Swell, the outcrop bands are considerably wider. In order, watch for dark red Triassic Moenkopi and Chinle formations covered in places with fine pink soil composed of windblown sand, massive Wingate sandstone with its sharp-edged ledges and cliffs, the Kayenta formation's slabby sandstone, and the dramatic white knobs and turrets of the Navajo sandstone, marked with sweeping cross-bedding.

Of the younger Jurassic formations farther west along the highway, the Entrada formation again appears in its racetrack valley, here ornamented with quaint "stone babies." The pale greenish sandstone of the Curtis formation forms a gently dipping cuesta near milepost 107. Tide-flat deposits of the Summerville formation can again be recognized by their castlelike bluffs and superabundant gypsum.

Dramatic vistas on the west side of the San Rafael Swell show, from top to bottom, the Navajo, Kayenta, and Wingate formations, major scenery-makers of Plateau-country Utah.

Near and west of milepost 114, gray gypsum carpets drape across some of the rocks. Given enough time and an adequate slope, gypsum flows like very thick taffy or syrup. There is so much gypsum in Jurassic rocks here that its flowing has in some places completely distorted the rocks, so that their regular layering has been destroyed. It even flows onto the surface to form broad carpets that very gradually move downslope.

Castle Valley and its badlands expose the weak but colorful Morrison formation, the uppermost Jurassic unit. The Dakota sandstone, the oldest Cretaceous formation, caps a west-dipping cuesta at the Muddy Creek bridge. Note the bumps in the highway pavement where it crosses the Mancos shale floor of Castle Valley. The formation is weak and incompetent, and contains clay minerals that swell when they get wet, so it forms a poor road base. In roadcuts, the coal layers of the Mancos shale appear to be thicker than on the east side of the anticline; some coal has been mined in this area. Channel-filling sandstone layers appear in some roadcuts, particularly near the Sevier-Emery county line. Scattered dark basalt boulders come from Fish Lake Plateau to the southwest.

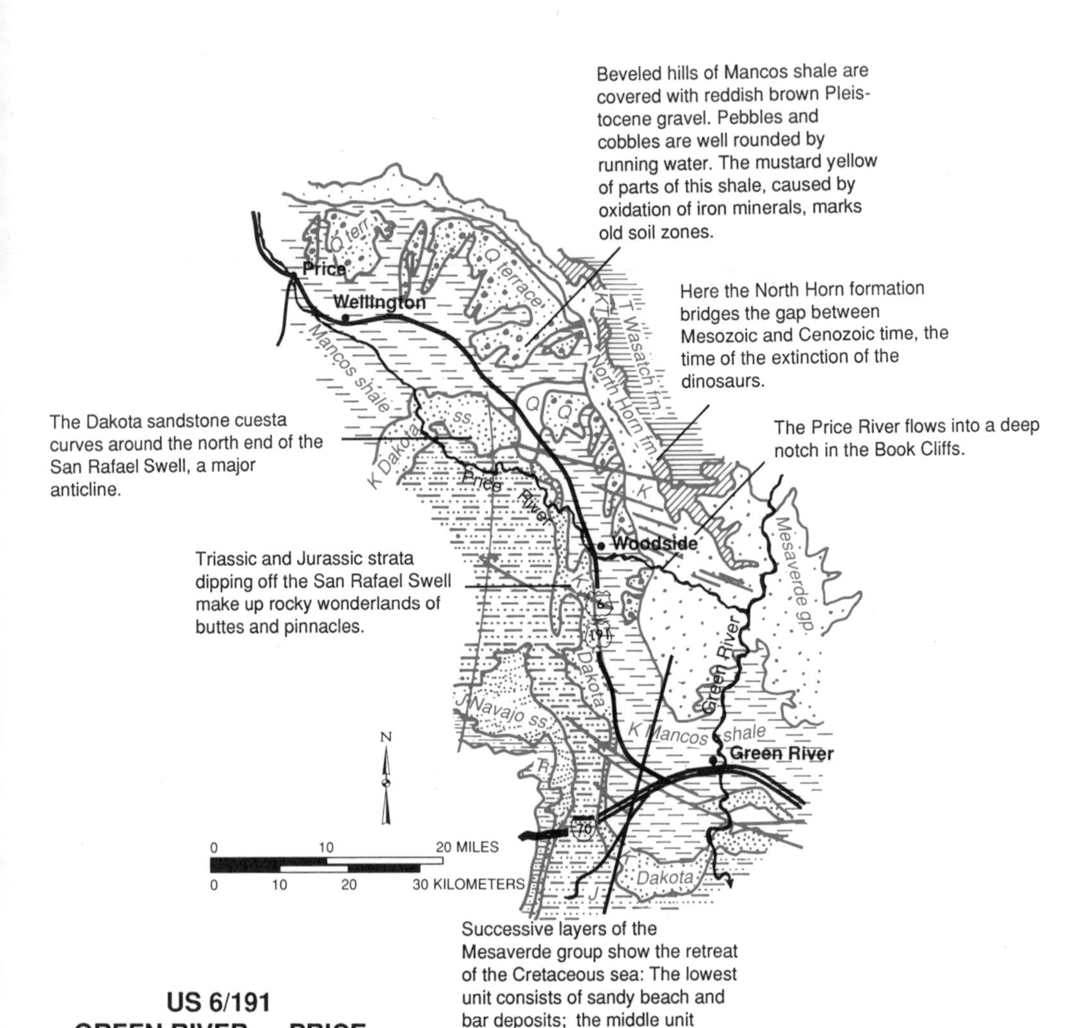

US 6/191
GREEN RIVER — PRICE

Above gentler slopes of Mancos shale, resistant sandstone ledges of the Mesaverde group form the Book Cliffs. The strata dip very gently north, and as erosion continues, the cliffs retreat in that direction.

US 6/191
Green River—Price
57 miles/92 km.

From its junction with Interstate 70, this highway heads northward into a desolate valley floored with Mancos shale, which will be with us all the way to Price. The broad anticline of the San Rafael Swell rises to the west, with steeply dipping Jurassic sandstone seeming to roll off its eastern flank. Near the highway, the Mancos shale includes thin coal beds formed from lagoon vegetation, as well as sandstone layers representing barrier bars to the east and stream deposits to the west.This area is near the western shore of the Cretaceous sea in which the Mancos shale was deposited. Some of the Mancos coal is mined northwest of Price, particularly in the area around East Carbon and Sunnyside.

To the east are the Book Cliffs, with Cretaceous rocks above the Mancos shale — the Mesaverde group. Still higher and out of sight from most parts of this highway are the Roan Cliffs, composed of

Tertiary (Paleocene and Eocene) lake deposits that accumulated in intermountain basins between the ranges of the newly risen Rocky Mountains.

At Woodside, carbon dioxide gas drives a geyser that erupts near the highway from a well originally drilled for water. When the well was drilled in 1910 the geyser spouted 75 feet high; it is much lower now. Eruptions occur about an hour apart. Other wells in this area also produce carbon dioxide, which is used for making dry ice. More carbon dioxide wells occur on the Farnham Dome north of the junction with Utah 123, near the mouth of Cat Canyon, and in and near Wellington.

North of Woodside the highway gradually curves westward around the north end of the San Rafael Swell. Badlands to the east culminate in ledges and slopes of the Book Cliffs and Roan Cliffs. The highway continues across gray Mancos shale, veneered with bouldery gravel brought from the cliffs during heavy rains.

North of here the Roan Cliffs include layers of sandstone impregnated with asphalt or tar, in beds 10 to 300 feet thick. These tar sands have been mined off and on for nearly a century, partly to use for asphalt pavement. This is the largest deposit of tar sand in the United States.

Price, on the Price River, lies near the western edge of the Plateau country, below the Book Cliffs and Roan Cliffs. Here the Green River formation of the Roan Cliffs contains another form of petroleum — oil shale. The oil shales, which extend east into Colorado, are a vast untapped petroleum reserve. The oil is so thick and waxy that the fine shale that contains it must be heated before it will separate and flow out — an expensive process at best. The deposits draw considerable attention each time oil prices rise.

The College of Eastern Utah Prehistoric Museum features reptile, mammal, and plant fossils found nearby, among them dinosaur skeletons and tracks. It also contains archeologic relics of a hunter-gatherer culture that populated this area more than 10,000 years ago.

US 40
Heber City—Duchesne
70 miles/113 km.

Starting within Utah's High Country, this route takes us across a geologically complex area where the Uinta Mountains intersect the Wasatch Range. That the Uinta uplift formed first is shown by the way its high anticline, faulted along both edges, continues west right through the Wasatch Range. There, ancient rocks in the heart of the anticline are exposed at the surface.

The Wasatch Range, steeply faulted on its western side, was once considered to be part of the Rocky Mountains, formed in late Cretaceous to early Tertiary time. Now it is known to be younger, and is considered the easternmost range of the Basin and Range region of Utah. That makes the Heber Valley the easternmost basin of the Basin and Range region. Deep below us as we cross the valley is one of the major thrust faults of the much older Sevier thrust belt, on which Paleozoic rocks moved east about 25 miles.

South of Heber City, the highway crosses the Provo River floodplain, skirting lava and breccia flows and thick layers of volcanic ash that arch around the western nose of the Uinta Mountains, partly covering Pennsylvanian and Permian strata. The volcanic rocks seem to have flowed, in Tertiary time, into the sag between the ranges. The Provo River drains the southeastern end of the Uintas, and is an important source of water for Utah's major cities.

East of the Provo River floodplain, US 40 heads up Daniels Canyon between massive ridges of dark gray Oquirrh group siltstone, sandstone, and limestone. These Pennsylvanian and Permian strata are part of the Sevier thrust belt, and are well east of their birthplace. They are sliced by many lesser faults, and alternate with similar fault slices of Triassic rocks, but vegetation and numerous rockslides and landslides make them hard to see.

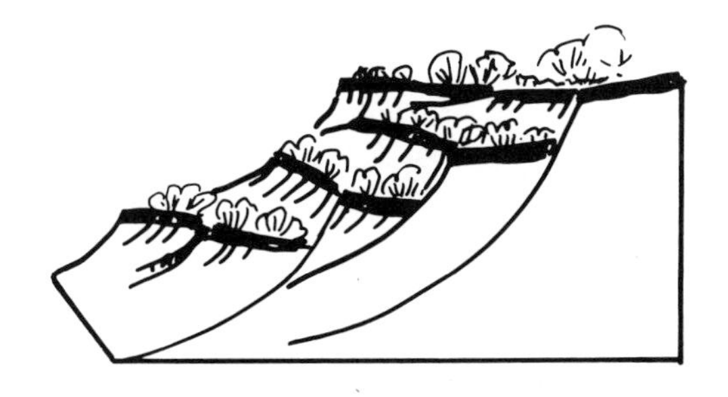

Small landslides and slumps, some with horseshoe-shaped scarps showing where they broke away from the slope, have damaged many highway cuts. In some of them there are several slide surfaces, so that the landslides appear as a number of little catsteps. Soil, grass, and shrubbery may remain intact.

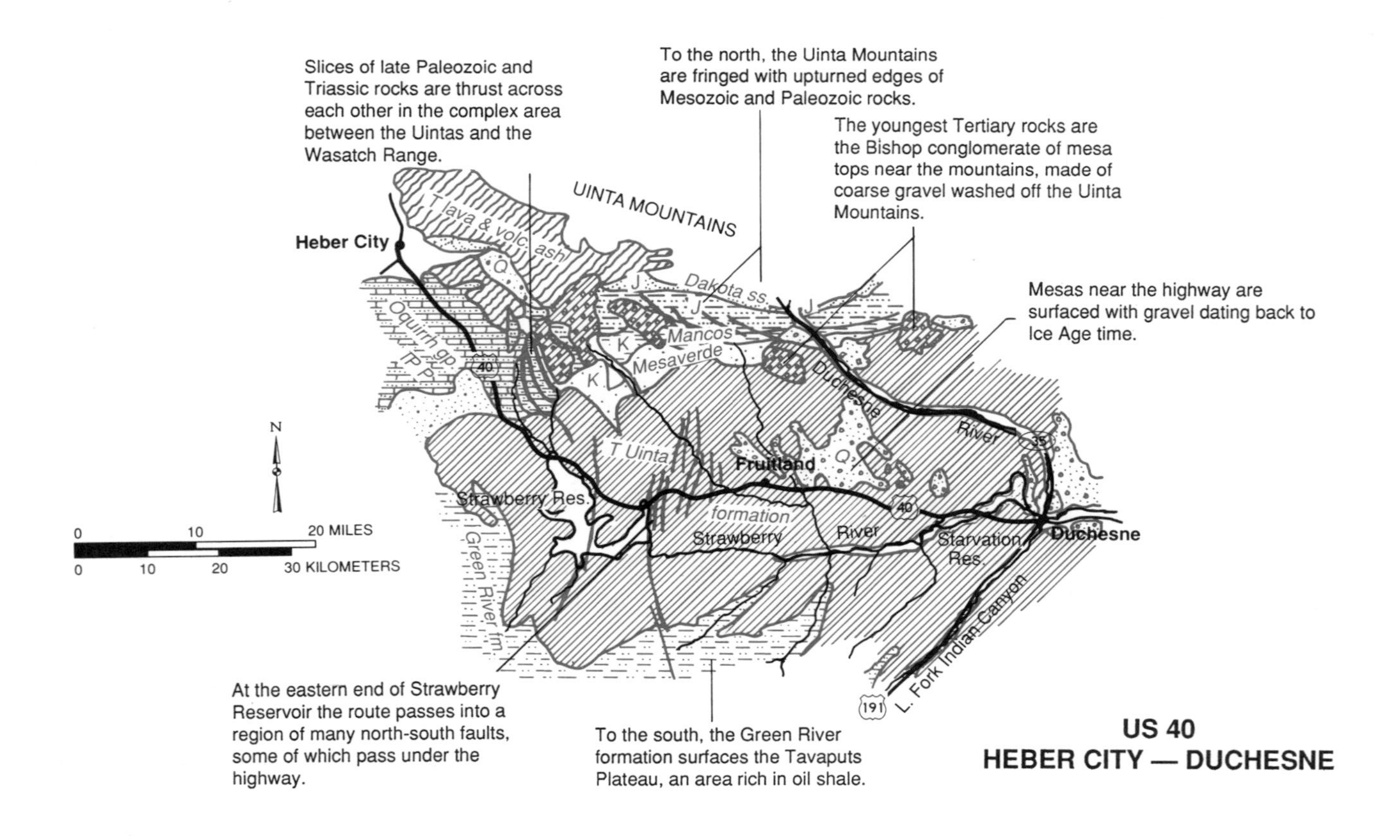

US 40
HEBER CITY — DUCHESNE

Water stored in Strawberry Reservoir flows west through a tunnel to serve heavily populated areas west of the Wasatch Mountains.

Sandstone saturated with tar occurs near the head of the canyon; at present, none is being mined. At the pass, at an elevation of 8000 feet, a light coating of glacial gravel covers the surface.

East of the pass the highway winds downhill toward Strawberry Reservoir, entering the Plateau country proper. The reservoir lies at the east portal of the tunnel that conveys eastern slope water through the mountains to Utah's cities west of the mountains. The Strawberry River originally drained east down the Duchesne River to the Green River and thence to the Colorado. Strawberry Reservoir dam, completed in 1916, reverses that flow: Water stored in it now goes west, not east.

Between Strawberry Reservoir and Duchesne the highway follows a surface of Tertiary rocks known as the Uinta formation — gravel, sand, and silt washed south off the Uinta Mountains. These Tertiary strata coarsen northward, toward their source in the mountains.

The Uinta Mountains, a long east-west anticline faulted along both edges, are part of the Rocky Mountains. They formed during Cretaceous-early Tertiary mountain building, as did the Rockies of Colorado and Wyoming. Their uplift can be accurately dated because late Cretaceous strata turn upward along their flanks, whereas the oldest Tertiary strata do not. Soon after they formed, they were virtually buried by sediments — mud and sand — washed off the mountains to the east and north. Then, during and after regional uplift in Miocene time, these soft sediments were eroded off their crest, exposing the hard rocks of the old anticline. Rocks in the high central part of the

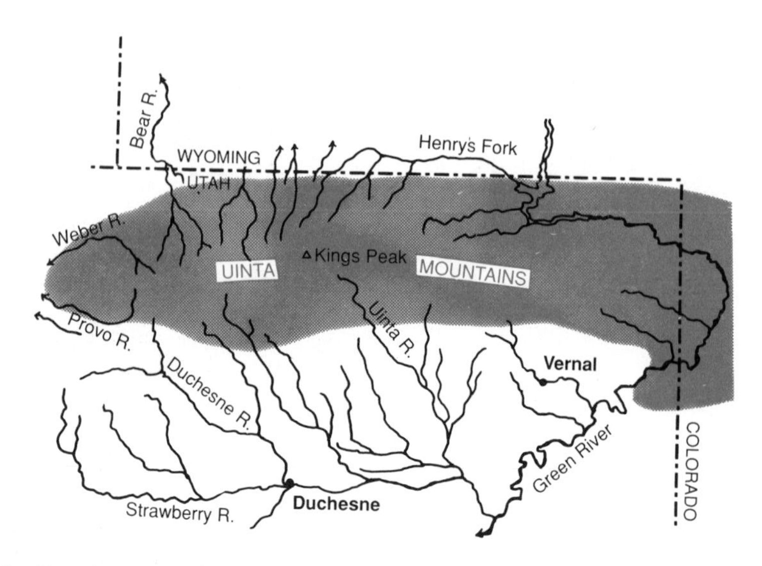

By diverting water from the Strawberry, Duchesne, and Provo rivers, Utah captures the abundant outflow from the south flank of the Uinta Mountains. The Bear River, after flowing north to Idaho, turns south again to empty into Great Salt Lake.

Uinta Mountains are mostly very ancient sedimentary rocks deposited during Precambrian time.

Mesas between the highway and the Uinta Mountains are capped with gravel and sand washed off the mountains in Pleistocene time. A few miles east of Fruitland we cross some of these long fingers of coarse, stream-deposited material. Southward, the Plateau country shows off its basically horizontal scenery.

The Uinta Mountains are surfaced on this side with Paleozoic rocks ranging in age from Cambrian to Pennsylvanian. Irregular, discontinuous ridges at the base of the mountains are upturned edges of Mesozoic rocks. Harder rocks form cuestas and hogbacks; softer units eroded into racetrack valleys that encircle the west end of the range, except where they are covered with lava flows.

The Uinta Basin south of the Uinta Mountains is a major oil province; numerous wells dot the area between Fruitland and Duchesne. The oil originates in Tertiary shale that is particularly rich in organic matter. It then migrates up into the Park City formation, a particularly porous limestone. There it is trapped by impervious red shales that are near-shore equivalents of the rock in which it originated. The natural gas in this field is also derived from decaying animal and plant material.

Starvation Reservoir, part of the Provo River Project, also collects water for cities west of the mountains. Though it may not feel like it, the Uinta Basin is so much higher than the Great Basin west of the mountains (7750 feet at Strawberry Reservoir and 5515 feet at Duchesne, vs 4300 feet at Salt Lake City) that the water flows by gravity through the tunnels, with enough energy to turn the turbines of generating plants on the way.

A view area near milepost 83 gives good views of the flat-topped Quaternary mesas northeast of Duchesne, stretching like long fingers from the mountains. Many of them are capped with glacial gravel. Note the two levels of river terraces near Duchesne.

Tan and reddish layers of sandstone and siltstone of the Uinta formation are well exposed in roadcuts near mileposts 57 and 58.

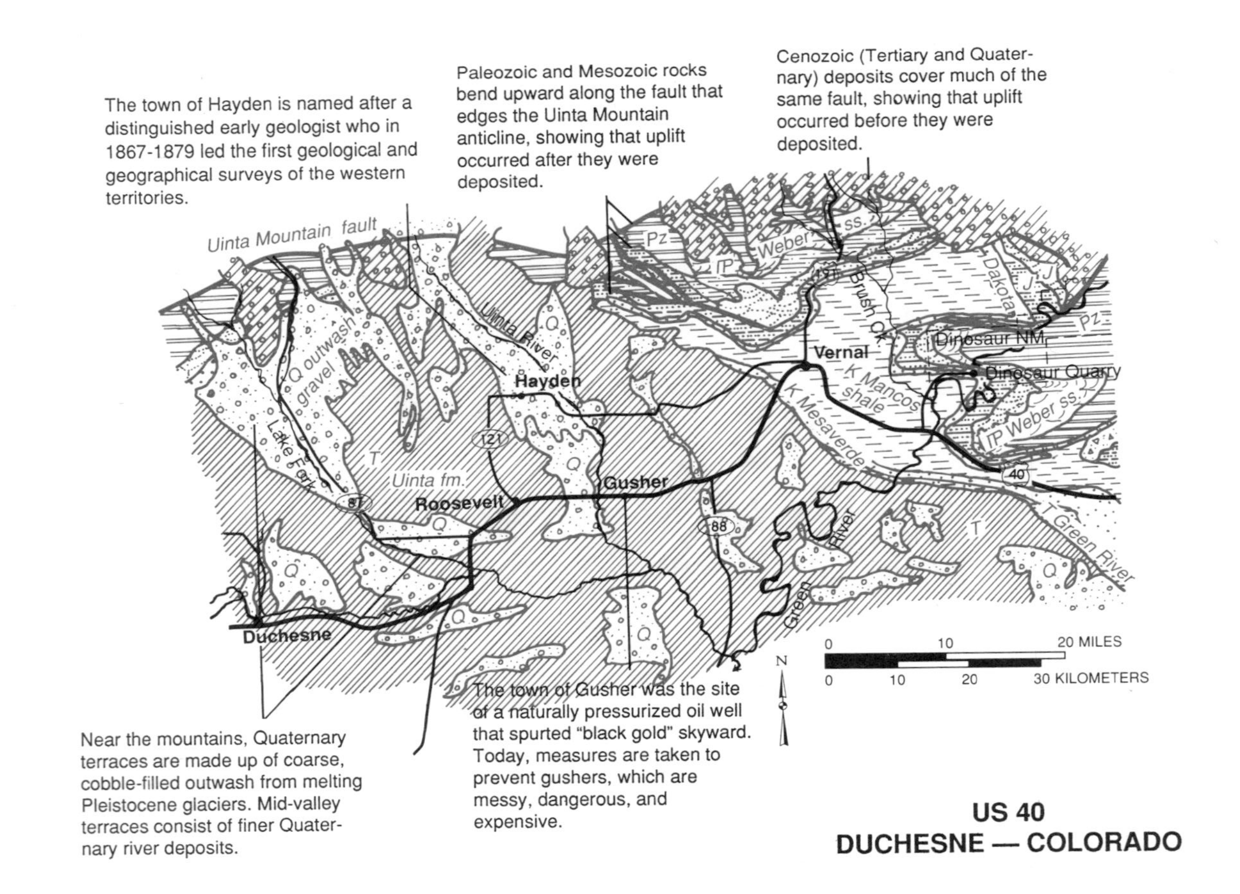

US 40
DUCHESNE — COLORADO

US 40
Duchesne — Colorado
88 miles/142 km.

At Duchesne we leave the valley of the Strawberry River and cross into the Duchesne River drainage. The Strawberry River, like the Duchesne, drains the south side of the Uinta Mountains. It was a tributary of the Duchesne until its flow was diverted, along with some Duchesne River water, into the Starvation and Strawberry reservoirs, to flow through tunnels to the west side of the mountains, where most of Utah's population lives and works.

East of Duchesne we get more of the wide-open feeling of the Plateau country, with flat-topped mesas a dominant feature of the scenery. Mesas near the highway are capped with Pleistocene gravel scoured in glacial times from Uinta Mountain summits to the north. The highway crosses the Duchesne River at Myton and ascends some of these mesas or terraces, with roadcuts exposing the cobble-filled gravels of which they are formed.

Born in Laramide time, as the Mesozoic Era ended and the Cenozoic Era began, the Uinta Mountains have shed a great amount of broken and stream-rounded rock material to the lower country that surrounds them. The range is basically a single large anticline about 160 miles long (east-west) and 30 to 40 miles wide, faulted along its north and south edges. Originally thousands of feet of Paleozoic and Mesozoic sedimentary rock must have covered the anticline; all that material has now washed down into surrounding basins, along with a sizeable amount cut by glaciers from the hard Precambrian mountain core.

The range is somewhat of an anomaly, oriented across the general north-south trend of the Rocky Mountains, of which it is a part. Radial drainage from the range collects into several sizable rivers: the Bear River, Henry's Fork, the Green River, and the Strawberry and Duchesne rivers, the last two now deprived of much of their water by diversion projects.

North and northeast of Roosevelt, fine pinkish sandstone and siltstone of the Tertiary Uinta formation show beneath the cap of Quaternary gravel. These rocks, formed from sediment washed off the Uintas, contain Oligocene fossils. Don't confuse the Tertiary Uinta formation here with the Precambrian Uinta Mountain group, a much older rock unit that makes up the heart of the Uinta Mountains.

A petroleum refinery near Roosevelt processes oil and gas from the oilfields of the Uinta Basin. Crude oil, as it comes from wells, is a blend of many different but related compounds, only a few of them useful without refining. Uinta Basin crude oil contains a lot of heavy oil that flows only when it is hot, so pipelines leading from the wells to the refinery must be heated.

In the refining process, large oil molecules are broken apart by high temperatures and pressures as well as by chemical cracking, then regrouped into useful products such as gasoline, diesel fuel, aviation and jet propulsion fuel, propane, butane, lubricating oils, kerosene, and various greases. Petroleum-derived chemicals also go into nylon, plastics, paints, detergents, photographic film, insulation, and hundreds of other products now in everyday use.

East of Roosevelt, Tertiary rocks dip toward the mountains, off a small anticline that more or less parallels the front of the mountains. This little anticline is the eastern end of the Greater Altamont Bluebell oil field. Anticlines, with their arching rocks, form traps in which oil and gas accumulate.

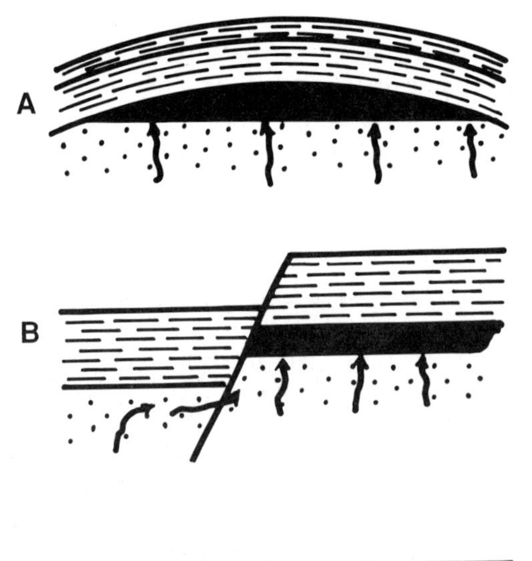

Oil in permeable rocks such as sandstone or cavernous limestone migrates upward until it is trapped in anticlines (A) or along faults (B) by impermeable rock — usually shale. Wedges of permeable rock (C) may also act as natural reservoirs.

Badlands develop in some of the soft Tertiary siltstones here, partly because of the nature of the rock, partly because of the arid climate. Some of the soft rocks contain volcanic ash, which makes the soil swell when it is wet and shrink when it dries, discouraging vegetation and highway-builders. Sudden but severe summer thunderstorms initiate and enlarge gullies in surfaces unprotected by plant cover. You can see the same effect on a smaller scale in highway cuts. The altered volcanic ash with its swelling and shrinking clays plays havoc with road surfaces; highway bumps are due not to the incompetence of highway engineers, but to incompetence of the clayey rock.

Watch highway cuts also for faults and small anticlines and synclines. Most of the pinkish rocks belong to the Uinta formation; they include massive cross-bedded sandstone and conglomerate in places, as well as weak mudstone layers that erode easily.

At Vernal, the Utah Field House Museum of Natural History, on US 40, emphasizes local dinosaurs. It is well worth a visit, whether or not you intend to visit the dinosaur quarry in Dinosaur National Monument, a short distance farther east.

East of Vernal a sharp, two-pronged anticline thrusts southward from the Uinta Mountains. Its north prong is Split Mountain, so-called because the Green River has cut a sheer-walled passage right through it. Geologists think the river established its course atop Tertiary deposits that at one time almost covered this part of the Uintas; the river detoured around what was then the exposed eastern end of the mountains. As the Tertiary deposits gradually eroded away, the river, carving ever deeper in the soft sediments, eventually

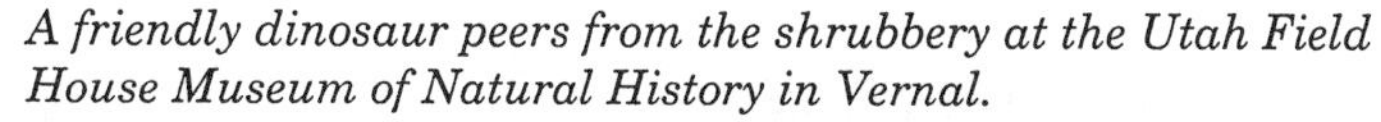

A friendly dinosaur peers from the shrubbery at the Utah Field House Museum of Natural History in Vernal.

Where the Green River emerges from Split Mountain, cross-bedded dune sandstone of the Permian Weber formation dips steeply southward off the Uinta Mountain uplift. Debris fallen from the cliff face is no longer removed by spring floods, thanks to Flaming Gorge Dam.

reached harder rocks below. Then, imprisoned in a canyon of its own making, the river retained its eastward swing and cut down right through Split Mountain. The dinosaur quarry lies at the southern edge of Split Mountain, near the point where the Green River emerges from its canyon.

Be sure to visit the dinosaur quarry and the mouth of the Green River's canyon. Hogbacks and cuestas along the base of the Uinta Mountains, visible from the route to the quarry and from Split Mountain Campground nearby, are the edges of Mesozoic and Paleozoic formations that swoop upward dramatically at the edge of the mountains.

The south flank of Split Mountain is surfaced with pale Pennsylvanian-Permian sandstone, the Weber formation. The sweeping crossbeds show that it was deposited in a field of sand dunes.

Eroded edges of Permian, Triassic, and Jurassic strata make a series of colorful chevrons resting against the Weber sandstone face of Split Mountain.

Ashley Valley oil field near Dinosaur National Monument produces oil from the Weber formation. This is the Uinta Basin's oldest oil field, discovered just after World War II.

Farther east along US 40, three terrace levels border the Green River. They formed in Pleistocene time, when the river's level and its load of gravel and sand changed with advancing and retreating ice. The highest terrace is the oldest.

Where the highway crosses the Green River at Jensen, roadcuts show that Cretaceous sedimentary rocks are beveled and covered with coarse Pleistocene gravel. The highway continues east on top of this gravel. Blue Mountain, the southern part of the Yampa Plateau, rises steeply, its face marked with chevrons of Permian and Triassic rock.

At milepost 164 we come out of the Cretaceous rocks and onto the Morrison formation, the colorful Jurassic rock that contains the dinosaur bones near Vernal. The Yampa Plateau, and the Uintas as a whole, continue eastward beyond the Colorado border, with Weber

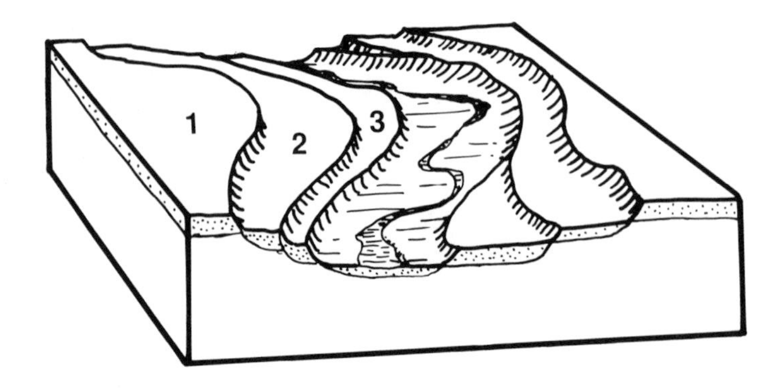

Successive river terraces develop during changes in river flow. Each terrace represents a former floodplain, deposited and then cut away by the shifting river.

sandstone surfacing their southern slope. A viewpoint near milepost 166 contains exhibits on the Uinta Mountains. Deep gullies nearby and in the last few miles before the Colorado border reflect a pattern of gullying seen all over the southwest: The gullies have developed since 1880, and reflect increased erosion probably due to grazing.

US 89
Arizona—Kanab
66 miles/106 km.

Leaving Arizona just north of Wahweap, this highway curves northwestward on a wide bench eroded in soft Jurassic rocks of the Carmel formation, above the pale cliffs of Navajo sandstone that wall Lake Powell and the Glen Canyon Dam site. Most of the mesas and buttes around Lake Powell top out in Carmel formation siltstones. Above the Carmel is the Entrada sandstone, which like the Navajo sandstone accumulated as Jurassic sand dunes. The two formations look somewhat alike, though here the Entrada sandstone is tan rather than pinkish like the Navajo sandstone.

Between mileposts 4 and 5, younger rocks appear to the north and northeast. The lowest yellowish brown ledges are Dakota sandstone, a Cretaceous rock unit, with a whitish slope or bench of Jurassic Morrison formation below. Above the Dakota sandstone are gray slopes of Tropic shale and light brown cliffs of the Straight Cliffs formation, both Cretaceous. They edge the Kaiparowits Plateau and together make up the Gray Cliffs of Utah's Grand Staircase. The plateau above them supports higher buttes of Tertiary sedimentary rocks — the Pink Cliffs, out of sight from this part of the highway.

Closely spaced vertical joints give some of these rocks a stockadelike appearance. As shale erodes and cliffs are undermined, blocks and slabs of sandstone break off along the joints and tumble down the shale slopes. With successive rockfalls the cliffs gradually retreat northward.

Patches of sand dunes appear along parts of this route. Their pale, even-grained sand, which at times drifts onto the highway, is recycled from much older dunes preserved in the Navajo and Entrada formations.

The highway has been climbing steadily since leaving Lake Powell. Near the bend in the road between mileposts 17 and 18, it crosses a drainage divide between streams draining toward Lake Powell and those draining toward the Paria River. Striped pink and gray rocks west of milepost 18 are a part of the Carmel formation, which includes many water-deposited siltstone and sandstone layers. The stripes are most pronounced near milepost 23.

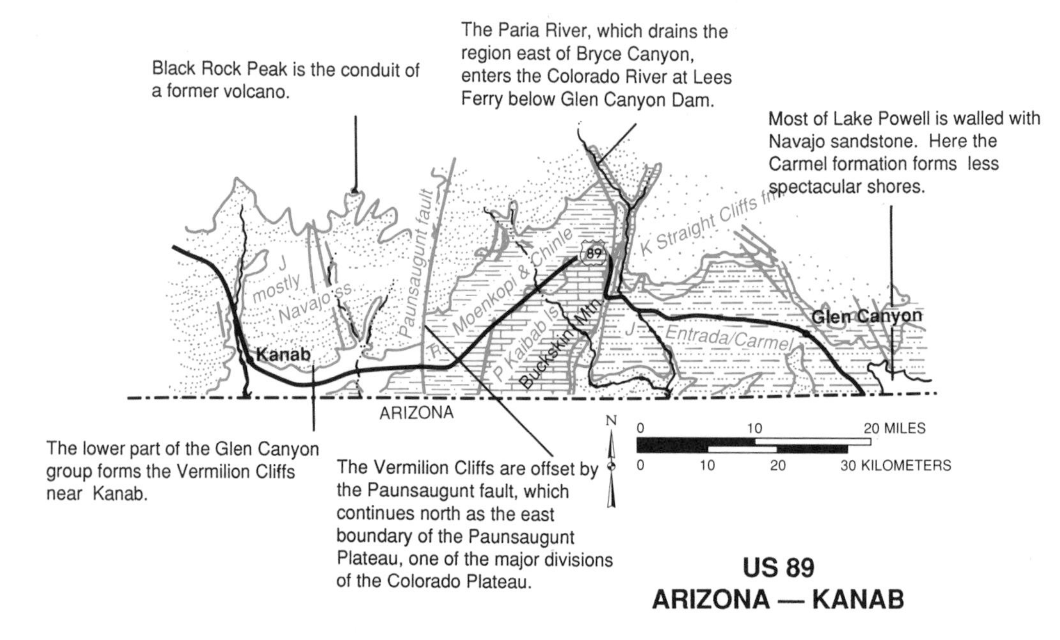

US 89
ARIZONA — KANAB

Candy-striped Jurassic mudstone and siltstone have eroded into fanciful badlands near the highway. The dry climate and occasional cloudbursts, as well as volcanic ash within the mudstone layers, prevent soil development here.

Highway cuts show that rocks upturned along the Cockscomb are offset by many small faults not visible in natural exposures.

Just west of milepost 21, where the highway crosses the Paria River, are good exposures of the Carmel formation in ledges to the north. An upper tongue of white Navajo sandstone also appears here. The Navajo sandstone is well exposed at the big roadcut between mileposts 24 and 25, where the highway slices through the steeply tilted strata. The route then crosses successively older rocks, with numerous roadcut exposures showing the colorful, steeply tilted rock layers.

At milepost 26 the highway runs due north along a valley eroded in tilted redbeds — red sandstone and siltstone — of the Triassic Moenkopi formation. The stream here has cut deep gullies in its older valley floor. Such gullying facilitates runoff and lowers the water table, ultimately expanding the area of desert.

Continuing northward along the valley we can easily see that Triassic and Jurassic rocks suddenly swoop upward, the most resistant among them forming a giant cockscomb that outlines the east flank of Buckskin Mountain. Parts of this mountain's surface are covered with windblown sand derived from the Navajo sandstone, which surfaces much of the area northwest of here. The erosion-resisting tan rock visible elsewhere on the mountain is the Kaibab limestone, the youngest Paleozoic formation here. This limestone also surfaces the Kaibab Plateau to the south, and both rims of Grand Canyon. To the north, red sandstone bluffs rise above gaily colored slopes of Chinle formation. As the highway curves around the north end of Buckskin Mountain it closely follows the line between the Kaibab formation and the red Triassic rocks of the Moenkopi formation.

As can be seen in some of the small canyons that cut into the flanks of Buckskin Mountain, the contact between the Moenkopi and Chinle formations is irregular, showing that there was a period of erosion after the Moenkopi formation was deposited.

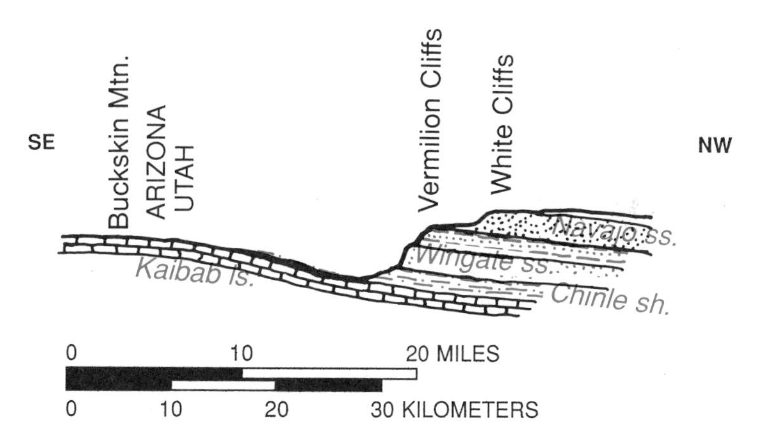

Section across Highway US 89 at Buckskin Mountain

From milepost 30 look north to distant cliffs of Tertiary sedimentary rocks, the Pink Cliffs of the Grand Staircase. Northwestward are several small peaks, one of them a grayish volcanic neck, the other two erosional remnants of Navajo sandstone raised high by the monocline of the Cockscomb. To the northwest are red cliffs of the same Triassic rocks we have seen in the Cockscomb, with colorful Chinle formation shales exposed on lower slopes. These rocks continue westward as the Vermilion Cliffs, which appear now in the long line of promontories guarding Utah's southern frontier.

Looking west from milepost 44, the Mt. Trumbull volcano (elevation 7700 feet) and several smaller volcanoes appear in the far distance.

In the valley between mileposts 45 and 46, the highway crosses a major fault, the Paunsaugunt fault. Rocks west of it dropped relative to those to the east. The effect near the highway is to bring the Vermilion Cliffs much closer, and to displace the Moenkopi formation southward, where it forms the Chocolate Cliffs. This gives us a good opportunity to see the red Triassic sandstone and siltstone of the Vermilion Cliffs and their thin pale cap of Navajo sandstone.

The Vermilion Cliffs, Kanab's backdrop, expose layers of red Triassic siltstone, mudstone, and shale capped with a thin white layer of Navajo sandstone. The cliffs weather and erode in the cliff-slope-cliff pattern characteristic of all of the Plateau country.

Here part of the Navajo sandstone again is interlayered with part of the Carmel formation, reflecting the advance and retreat of dunes of the Jurassic desert. Farms and ranches in this region obtain their water from the lower tongues of the Navajo sandstone. Recently, coal companies planning to mine Cretaceous coal farther north on the Kaiparowits Plateau announced their intention to drill into the Navajo sandstone for water — enough of it to mix with ground-up coal to make a slurry which can be pumped through a pipeline to Nevada power plants. Though they plan to drill only to the upper part of the Navajo sandstone, farmers here are understandably fearful of losing their water. White cliffs of Navajo sandstone can be seen to the north from milepost 78.

Near Kanab the highway runs on the lower part of the Chinle formation, with thin but resistant Shinarump conglomerate at the base of the formation capping a prominent ridge to the south. All the rocks dip very gently northward. Above the town, the whitish lower tongue of the Navajo sandstone can be seen interlayered with the red Carmel formation.

As the highway approaches Kanab, purple hills of the Chinle formation appear close by. The town lies on Kanab Creek at the base of the Vermilion Cliffs.

US 163
Arizona—Bluff
45 miles/72 km.

This route enters Utah in the western part of Monument Valley. Here on the crest of a wide anticline, the Monument Upwarp, Pennsylvanian and Permian rocks combine to form some of Utah's most spectacular scenery, early recognized by Hollywood as a stunning background for western films.

Monument Valley is within the Navajo Reservation, and narrow dirt roads are closed to traffic except with Navajo guides. However, a graded road leads east from US 163 to an overlook that affords good views.

The buttes and pinnacles of Monument Valley are composed of Cedar Mesa sandstone. The slopes at their bases are Halgaito shale. Some of the monuments are capped with red ledges of Organ Rock shale.

The highway near the state line crosses a slightly domed surface of hard, resistant limestone belonging to the Honaker Trail formation, generally covered by windblown sand. The uppermost of the Pennsylvanian rocks in this area, this formation can be seen to better advantage near Monument Pass, where the dune sand is less abundant. A near-shore marine limestone, the formation contains fossils of Pennsylvanian bryozoans, tiny foraminifera, and brachiopods.

Slopes below the tall cliffs of the monuments expose reddish brown Permian shale and thin ledges of sandstone. Sheer cliffs of the monu-

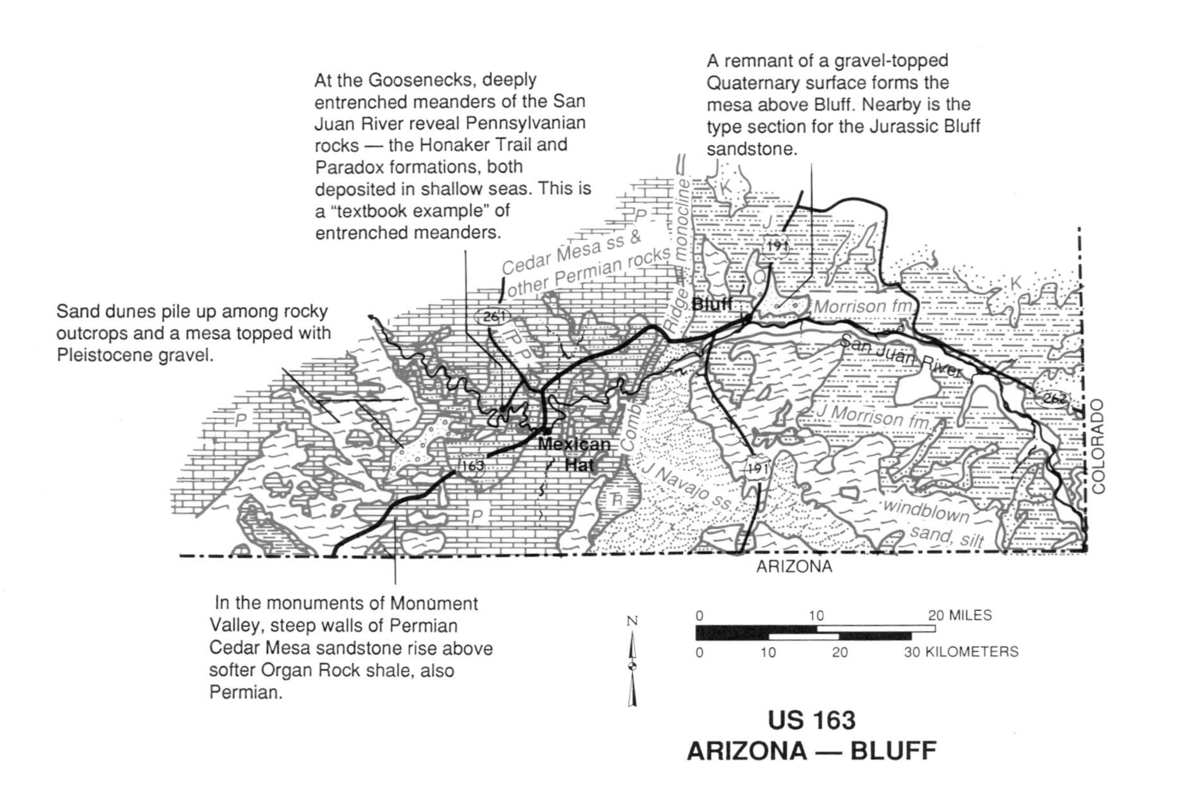

US 163
ARIZONA — BLUFF

Eroded by wind and rain, soft red shales undermine strong, vertically jointed sandstone to produce many free-standing buttes and pinnacles.

ments are made of another Permian formation, the Cedar Mesa sandstone. This dune sandstone, like many others, tends to break along vertical joints. Some of the monuments wear layered caps of Organ Rock shale; on a few, red Triassic mudstone forms the highest summits.

Drifting sand dunes along the highway recycle sand from dunes originally deposited 300 million years ago!

East of the viewpoint for Alhambra Rock, hogbacks of Comb Ridge mark the east side of the Monument Upwarp. Their rocks are mostly Jurassic, younger than those of Monument Valley itself.

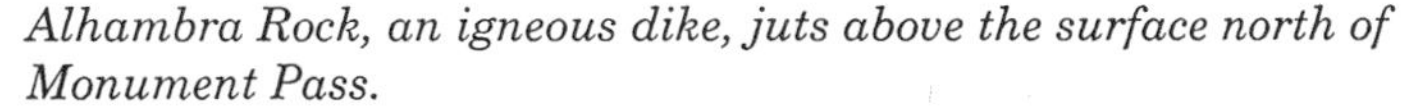

Alhambra Rock, an igneous dike, juts above the surface north of Monument Pass.

Limestone of the Honaker Trail formation is well exposed near the San Juan River bridge. Though harnessed by upstream dams, the river flows well enough in spring and early summer to attract river rafters. In fall and winter it may be almost dry.

The San Juan River rises in the San Juan Mountains of southwest Colorado. Before upstream dams were built, the San Juan carried an average 34 million tons of sediment per year to the Colorado River. During a 1911 flood the river rose 50 feet above its low-water mark in canyons downstream from Mexican Hat.

The Mexican Hat that gives the settlement north of the river its name can be seen from the town, on the skyline to the northeast. A

small oilfield near here, discovered in 1908, brought a few years of prosperity to the little community. Oil was produced from limestone reefs in Pennsylvanian rocks. It is unusual to find oil in a syncline, as here, rather than in an anticline. Uranium has been mined here as well; an old uranium plant near the town once processed ores from this region.

Four miles north of Mexican Hat, Utah 61 leads west to the Goosenecks of the San Juan, well worth a visit. Portraits of these goosenecks appear in many geology textbooks as the classic example of incised meanders, where a river that bends and loops lazily across a scarcely sloping surface is given new life by uplift, and cuts downward while still retaining its meandering course.

There are a number of interesting hoodoos near the highway in the area north of Mexican Hat, their weird figures formed by differential erosion of hard and soft layers in Permian rocks. Another side road goes to Valley of the Gods Scenic Area, about 14 miles of dirt road, where the Cedar Mesa sandstone, the same rock that forms monuments in Monument Valley, has been shaped by erosion into rows of statuesque figures.

A few miles farther northeast this route climbs across the Limestone Ridge anticline. Marine limestone of the Honaker Trail formation surfaces the anticline. Roadcuts show that many individual layers within this formation thicken and thin, a feature probably due to reeflike accumulations of lime-secreting algae and animal shells.

The Goosenecks of the San Juan are textbook examples of incised or entrenched meanders. Upper slopes of the 1000-foot gorge expose the Honaker Trail formation; lower cliffs are part of the Paradox formation.

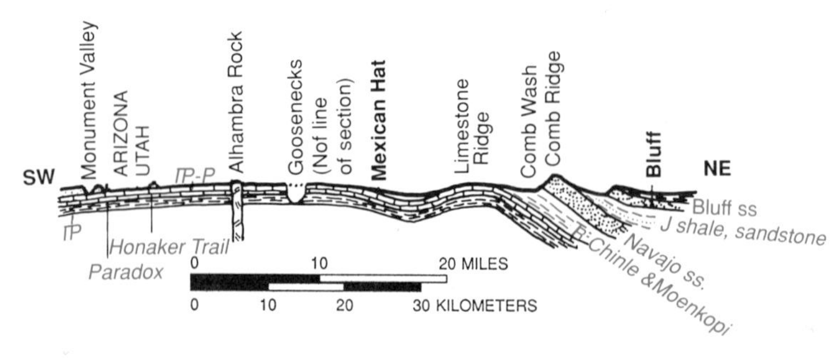

Section along US 163 Arizona-Bluff

Descending from Limestone Ridge anticline, the highway encounters Permian rocks again — tilted red mudstone that lies on Pennsylvanian limestone, and then the Cedar Mesa sandstone. Where the highway descends to cross Comb Wash, the Cedar Mesa sandstone includes a massive bed of gray gypsum.

At Comb Wash, the highway crosses a long, straight valley eroded in soft but colorful Triassic rocks. The succession of rock layers here, forming long lines of cuestas and hogbacks, provides an excellent example of large-scale differential weathering and erosion. Forming the crest of Comb Ridge is the Navajo sandstone, the sandy accumulation of a Jurassic Sahara. Its sweeping cross-bedding records the lee slopes of countless individual sand dunes.

Hogbacks of Comb Ridge edge the Monument Upwarp between Limestone Ridge and Bluff. The crest of the ridge is in Navajo sandstone; weak Triassic rocks form the "racetrack" valley in the foreground.

Younger rocks appear east of Comb Ridge: the reddish Carmel formation, and still farther east the light-colored, cross-bedded Bluff sandstone. Both are Jurassic. The Bluff sandstone, forming cliffs both north and south of the San Juan River, displays plenty of dune cross-bedding but represents a much smaller dune region than does the Navajo sandstone.

Approaching Bluff the highway descends across successive flat-topped, gravel-covered terraces deposited by Cottonwood Wash and the San Juan River. The terraces formed as the two streams deposited and then cut down through several successive floodplains. The San Juan is the major tributary of the Colorado River in this area. Along this relatively peaceful stretch it carves a passage through soft red siltstone, undermining the massive Bluff sandstone above.

Opinions differ on the nature of the Abajo Mountains. Some geologists say they are stocks, intrusions that have no real base. Others say they are laccoliths, intrusions that squeeze between layers of sedimentary rock, doming those above.

Ledges and small cliffs near the highway as it crosses Recapture Creek and Devils Canyon are typical Dakota sandstone, which commonly weathers into rectangular blocks.

The Comb Ridge monocline extends from Kayenta, Arizona, east and north to the Abajo Mountains, a distance of about 100 miles. It forms the eastern flank of the Monument Upwarp.

The Navajo Twins and Sunbonnet Rock are good examples of differential weathering of hard and soft rock layers.

The Bluff sandstone, limited in extent to the area north and south of Bluff, forms bluffs on both sides of the San Juan River.

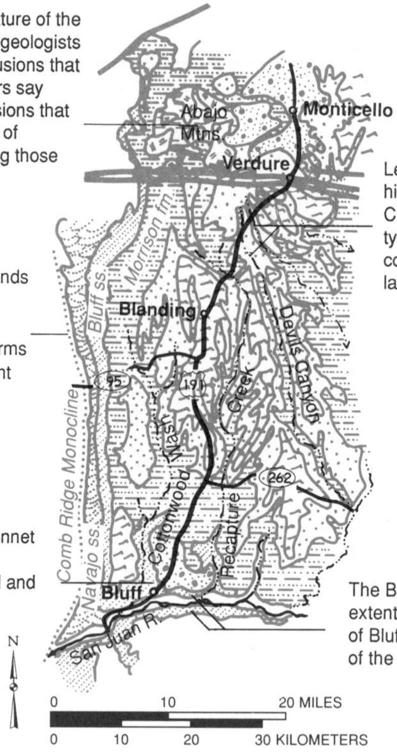

Abajo Mtns. — Monticello — Verdure — Blanding — Bluff — San Juan R. — Bluff ss. — Morrison fm — Comb Ridge Monocline — Navajo ss. — Cottonwood Wash — Recapture Creek — Devils Canyon — 95 — 191 — 262 — N

0 10 20 MILES

0 10 20 30 KILOMETERS

US 191
BLUFF — MONTICELLO

Erosion along vertical joints in the Summerville and Bluff formations created the Navajo Twins. The crinkly, irregular mudstone layers of the Summerville formation were distorted, while they were still quite soft, by the weight of overlying sand. Sunbonnet Rock is in the foreground.

US 191
Bluff—Monticello
48 miles/77 km.

Bluff lies at the confluence of Cottonwood Wash, usually dry or just barely flowing, and the San Juan River. The wash is typical of streams in the arid southwest: Overloaded with sand and gravel, its flow shifts with every storm, leaving braided patterns in its otherwise flat, sandy bed. The San Juan's channels shift as well, but on a longer time-scale, with gravelly islands that may persist for many years. Upstream, the San Juan, with headwaters in the Rocky Mountains, is impounded behind several dams; though it still flows abundantly in spring and early summer, it may be nearly dry at other times of year.

From 1891 to 1893 Bluff saw a mini-gold rush when some 1200 prospectors descended on this stretch of the San Juan River. Sand and gravel along the river were washed for placer gold. One operation produced $3000 worth of gold in one month — quite a fortune in the 1890s. By and large, though, most of the gold was too finely disseminated for profitable mining.

The Navajo Twins, just north of Bluff on the old highway, were carved by erosion in soft rocks of the Summerville formation, capped by more resistant Bluff sandstone. Sunbonnet Rock is a fallen block of Bluff sandstone that similarly protects softer rock beneath.

Much of the erosion in this area, particularly of features like the Navajo Twins and Sunbonnet Rock, is the work of wind as well as of water. Wind erosion is strongest near the ground, where gusts pick up particles of sand and hurl them against nearby rock. However, both wind and rain work slowly: The Navajo Twins were photographed in 1875 by William Henry Jackson, photographer for an early government survey of the western territories. At the time, geologists on the survey team were sure the Twins would soon fall. But comparison of today's scene with the Jackson photographs shows little change. Erosion in this region proceeds at an average rate estimated at about 1/4 inch per 100 years.

North of the Navajo Twins the route climbs through a narrow canyon cut in the Summerville and Bluff formations, to the bench at the top of the Bluff sandstone. Fine windblown silt covers part of this surface. The highway remains on this bench for some distance. To the northeast are outlying mesas of pink and yellowish gray rocks of the Morrison formation. The pale gray or white Navajo sandstone appears to the west, where it swoops upward to form jagged-edged Comb Ridge. Farther west are forested slopes of the Monument Upwarp, of which Comb Ridge is the steep eastern edge.

The high country north of Bluff overlooks much of the Four-Corners region, the only place in United States where four states (Arizona, Colorado, New Mexico, and Utah) meet. The San Juan Mountains of

The Abajo Mountains formed as several small laccoliths pushed upward, doming overlying sedimentary rock layers. Remnants of the domed layers surround the mountains and appear between the present peaks.

The Abajo Mountains show a characteristic of many southwestern mountains: Vegetation grows thickest on cooler, moister, north-facing slopes, to the right in this photograph.

Colorado are visible along the northeastern skyline; Sleeping Ute Mountain, shaped like a blanket-covered Indian with his toes sticking out, appears to the east. To the north are the Abajo Mountains. Both of these islandlike ranges are clusters of small igneous intrusions that lifted and domed overlying layers of sedimentary rocks during Tertiary time. There are other such clusters, called laccoliths, in southeastern Utah: the La Sal Mountains near Moab, the Henry Mountains southwest of Hanksville, and Navajo Mountain south of Lake Powell. Southeast of the viewpont are the mesas and tablelands that surround the Four Corners.

Near the junction with the Hovenweep National Monument road are good exposures of rainbow-colored Morrison formation, which also shows up on the bluff along the south edge of White Mesa, just north of the junction. This formation was deposited in Jurassic time as the fine mud and clay of river floodplains, with a generous lacing of added volcanic ash. The floodplains were crossed by winding, sand-filled channels that are now visible as lenses of resistant sandstone protruding from the weaker mudstone.

The highway climbs to the top of White Mesa, which is capped with hard, resistant Dakota sandstone, a wide-ranging near-shore deposit that records the invasion of the Cretaceous sea — the last to have entered Utah. The road remains on the Dakota sandstone for some distance. The irregular contact between the drab Dakota sandstone and the colorful Morrison formation below it shows well in big roadcuts north of Blanding.

On White Mesa, around Blanding, and around Monticello reddish soils derived from windblown silt and dust conceal the Dakota sandstone. Thanks to these fine, uniform soils, the mesa surfaces support farming; the area was once famous for its pinto beans. Under the soils, the Dakota sandstone can be seen in the walls of Recapture Creek and Devils Canyon. Bright green rocks exposed in some roadcuts contain glauconite, a mineral that forms in sea water.

Near the little community of Verdure the highway crosses a graben, a narrow slice of land where the Dakota sandstone and rocks beneath it have dropped down about 200 feet between two parallel east-west faults.

In Montezuma Canyon southeast of Monticello is an abandoned uranium mill. The Monticello uranium district was for a time Utah's most productive. The ore came from channel sandstones in the lower part of the Morrison formation, and from the Triassic Chinle formation. Since uranium tends to precipitate where plant debris has accumulated along stream channels, deposits in this region tend to be long, slender, and winding. Prehistoric Indian dwellings have also been found in Montezuma Canyon.

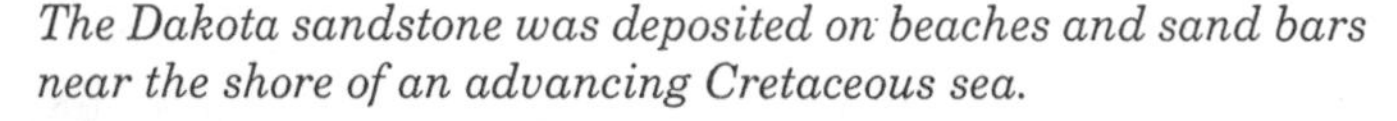

The Dakota sandstone was deposited on beaches and sand bars near the shore of an advancing Cretaceous sea.

Three divisions of the Entrada formation appear on Church Rock: The uppermost Moab tongue caps the thick, massive Slickrock member, with the Dewey Bridge member as the base. Background mountains are the La Sals.

US 191
Monticello — I-70
86 miles/138 km.

North of Monticello and its now inactive uranium mill, the highway runs on a gravel surface, with pebbles and cobbles of igneous rock washed from the nearby Abajo Mountains. Under the gravel is the Mancos shale, deposited in a Cretaceous sea, poorly consolidated and tending to form a weak roadbed — hence the bumpy highway. Below the Mancos shale, visible in deep gullies and canyons, is the Dakota sandstone, formed on beaches and bars as the Cretaceous sea flooded in from the east.

About ten miles north of town, castle-like outcrops along the west edge of the valley are fine Jurassic siltstone, the Summerville formation. Watch for small faults that offset hard and soft bands of this formation.

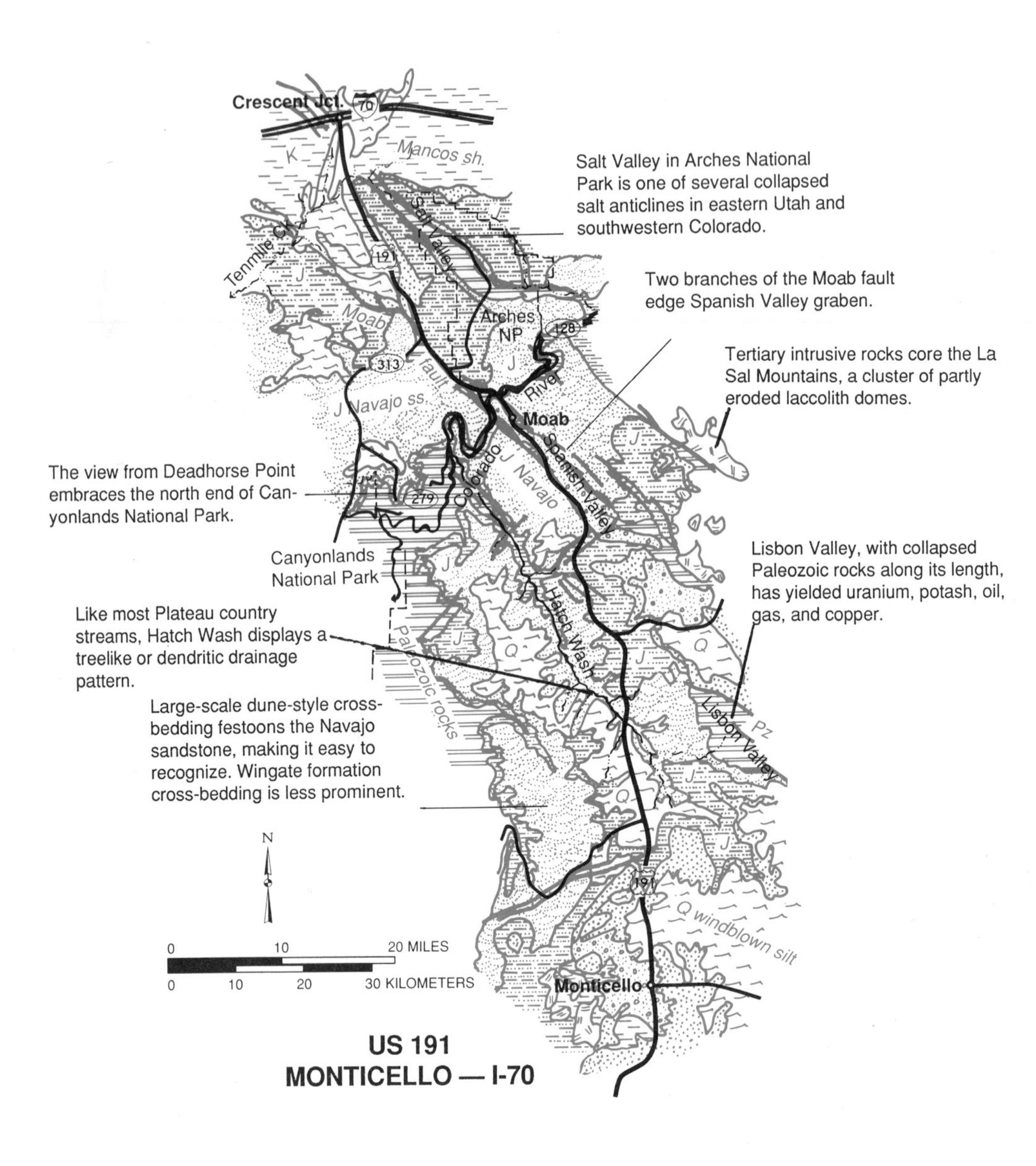

US 191
MONTICELLO — I-70

Church Rock, east of the highway near milepost 87, is a prominent erosional remnant of the Entrada formation, a normally resistant rock unit that contributes to the splendid scenery in and around Canyonlands National Park. Here as elsewhere an upper massive sandstone of the Entrada formation is undermined as lower mudstone layers are worn away.

The junction near Church Rock leads to the southern part of Canyonlands National Park, described in Chapter IV.

Lisbon Valley, east of this highway, is an unusual type of anticline caused by flow of salt, gypsum, and potash deep underground. These minerals, called evaporites, accumulated where sea water evaporated in restricted bays during Pennsylvanian time. They were later covered with sedimentary rock layers.

Given time and adequate pressure, salt flows like glacial ice — very slowly, without becoming truly liquid. Less dense than overlying rocks, it flows from places where pressure of overlying rocks is great to places where pressure is less great — in this area along fault ridges of Precambrian rocks. Under the pressure of overlying rocks, the salt (using the word to include potash and gypsum as well as common "table" salt) moved slowly toward the old fault ridges, which trend roughly northwest, pushing overlying rocks upward — here in anti-

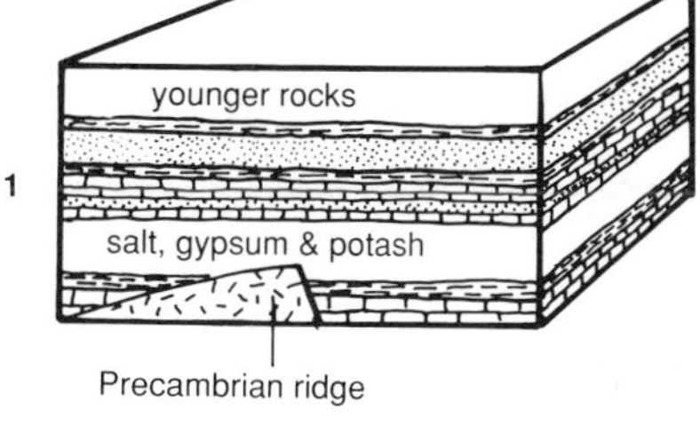

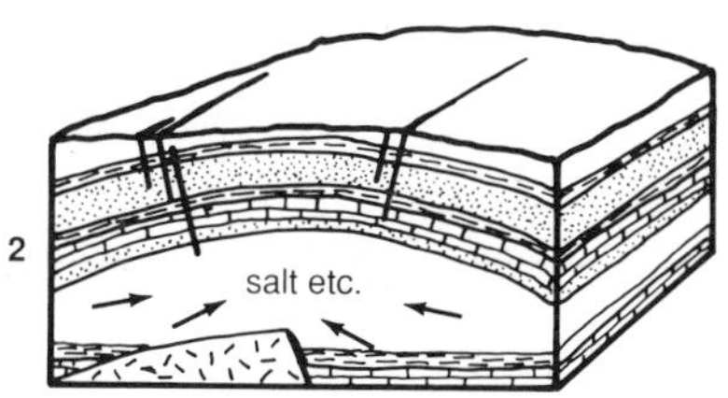

As salt layers in Pennsylvanian rocks pushed upward, Lisbon Valley anticline and similar salt anticlines developed in eastern Utah and western Colorado. Later, solution of salt caused the anticlines to collapse.

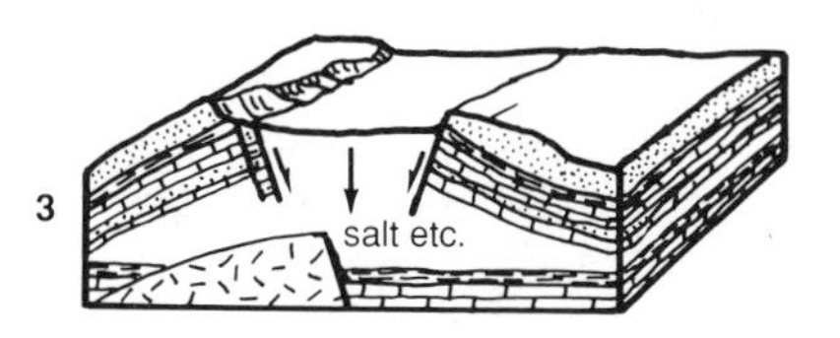

clines, there along faults. Several salt anticlines exist here. In some, like the Lisbon Valley salt anticline, groundwater gradually dissolved out the soluble salt, causing the anticline to collapse.

Several of the salt anticlines in this region serve as traps for upward-migrating oil and gas. Oil wells in Lisbon Valley produce several petroleum products, including propane, butane, and natural gasoline. Some of the gas is reinjected into the wells to force more oil to the surface. Also from this valley comes 10% of U.S. uranium production, most of it, along with vanadium, from the Triassic Chinle formation. And copper is mined nearby along the faults that edge the Lisbon Valley salt anticline.

A number of side roads lead into Canyonlands National Park, including those to Indian Creek State Park and Canyon Rim Recreation Area, and Needles and Anticline overlooks. The remarkable arches of Canyonlands and Arches national parks, both discussed more fully in Chapter IV, take shape because of differences in erosion patterns of two units of the Entrada sandstone — a thick, massive sandstone layer and, below it, a more readily eroded layer of crinkled, dark red mudstone. Erosion of the mudstone undermines the resistant sandstone above, and portions of the sandstone then fall away along curving joints. Though arches are concentrated in the national park areas, a few, such as Wilson and Window arches, are near this highway.

The area around Window Arch produces uranium and vanadium from the Morrison formation, a Jurassic unit readily recognized by its pastel colors. This region also contains abundant natural gas. A gas

Wilson Arch south of Moab is shaped in massive Entrada sandstone. The arch formed as a narrow fin of rock was eroded from both sides by wind, rain, and frost.

Barren cliffs of Entrada sandstone are topped with a thin layer of mudstone. In the background, the La Sal Mountains dominate the landscape. Notice the prominent vertical joint directly below the arrow. Minerals in groundwater seeping along this joint have hardened adjacent rock, which stands in a pinnacle and supports a prominent bulge in the cliff behind it.

compressor station five miles north of Window Arch links with a 26-inch pipeline to Colorado, New Mexico, and other western states.

The La Sal Mountains, in view to the northeast, are laccoliths formed in Tertiary time as molten igneous rock pushed upward, doming overlying sedimentary layers. The sedimentary rocks have long since been eroded from their summits, laying bare the intrusive igneous rock of the laccoliths — a gray, coarsely crystallized rock called diorite. The Abajo Mountains near Monticello and the Henry Mountains near Hanksville have similar histories.

The highway continues north, approaching Moab via Spanish Valley, another collapsed salt anticline edged by two branches of the Moab fault. Rounded orangish knobs of Jurassic Navajo sandstone appear west of the highway, above Triassic sandstone and shale and some whitish hills of gypsum that, like salt, flowed up or squeezed up along the fault. East of the highway, on the other side of the valley, the Navajo sandstone forms more regular bluffs, with collapsed rubble of other formations along their base. An old Spanish trail from Santa Fe gives this valley its name.

The Entrada sandstone, also formed on Jurassic dunes, rises above the Navajo sandstone, forming higher bluffs. This formation is an especially good water-carrier, an aquifer. Some of its water comes to the surface at Cane Springs State Park, where the rocks rose close to road level along a fault. The fault can be seen about half a mile north of the park area, where less permeable rocks abut the cliff-forming Entrada sandstone.

The Moab area is particularly interesting geologically. Rocks are all mixed up, both because of movement in gypsum and salt beds far below, and because of collapse of the Spanish Valley salt anticline. About a mile south of Moab, dirty gray gypsum can be seen through trees to the southwest, where it broke through overlying units. Similar features where rising gypsum has broken through overlying rock exist on both sides of the Colorado River's gorge near Moab; they can be seen from Utah 128 upriver from the town. Potash deposited with the gypsum and salt is mined and concentrated a few miles downriver from Moab. The Moab area was home base for Charlie Steen, whose 1952 uranium discoveries made him an instant millionaire. His Mi Vida mine in Lisbon Valley produced uranium from a channel sandstone 3000 feet long and 800 feet wide — the largest deposit in this area. All of the rocks visible from the town are Mesozoic — the same rocks you see in Canyonlands and Arches national parks. But Paleozoic rocks like the Paradox formation with its several

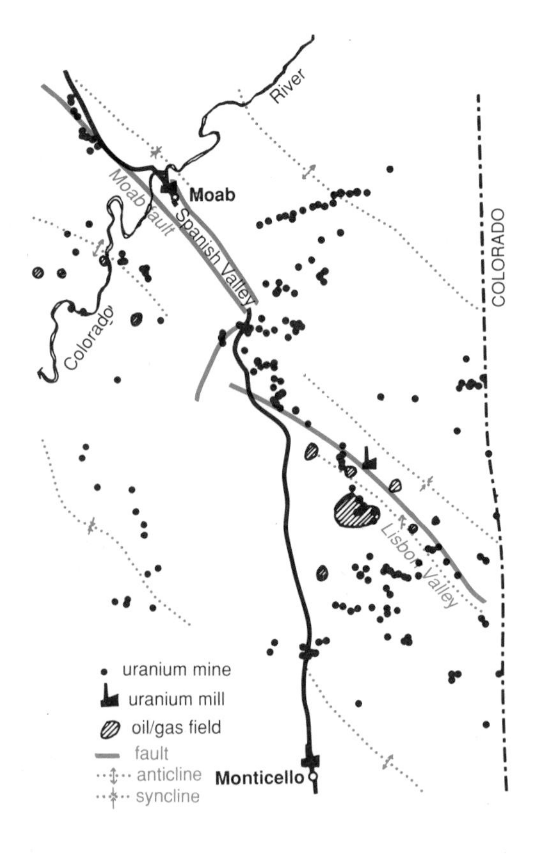

Uranium mines, most of them inactive now, dot the region between Monticello and Moab. Oil and gas are produced from this area, too.

thousand feet of salt and gypsum also played a significant part in creation of this scenery.

Just north of Moab is a large uranium mill formerly operated by Atlas Minerals. It can process 1000 tons of ore per day.

Leaving Moab the highway follows the Moab fault, the northwestward extension of the two faults edging Spanish Valley. The two branches of the fault come together here, so the valley is narrower than its counterpart southeast of Moab. West of the highway, the rocks are Pennsylvanian to Jurassic, with the Navajo sandstone along the skyline. East of the highway, the rocks are all Jurassic with the Navajo sandstone below the surface and the Entrada formation on the skyline. Total displacement along the Moab fault is 2000 feet. Test wells revealed about 7900 feet of underground salt, gypsum, and potash. What immense amounts of sea water must have evaporated to leave this much solid matter behind!

About eight miles north of Moab, outcrops of soft but colorful Jurassic rock of the Morrison formation appear west of the highway. This formation is rich in uranium, and wherever it is exposed it is dotted with uranium prospect pits and small mines.

A side road leads to Dead Horse Point and the northern parts of Canyonlands National Park — a worthwhile side trip offering a spectacular view of the redrock country entrenched by the Colorado and Green rivers 2000 feet below.

Near this junction, highway US 191 leaves the Moab fault zone, which continues northwest for another 40 miles. More green Jurassic rocks of the Morrison formation are exposed. As elsewhere, the

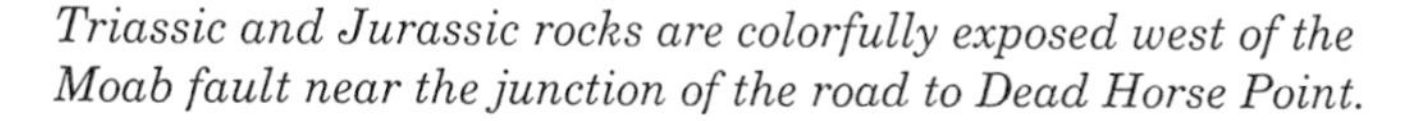

Triassic and Jurassic rocks are colorfully exposed west of the Moab fault near the junction of the road to Dead Horse Point.

Morrison formation here contains many small prospect pits and a few old uranium mines, most of them now inactive. Prominent bluffs above expose the Entrada formation, the arch-former of Arches National Park.

Farther north, near the microwave relay station, we can look southeast into the heart of the Salt Valley anticline, the prominent fold responsible for most of the arches of Arches National Park.

North of Rockhouse Creek the Dakota sandstone is exposed in bluffs east of the highway. Gem-quality red silicified dinosaur bone fragments derived from the Morrison formation, as well as barite nodules, can be found in stream gravels near the Valley City Reservoir five miles south of Interstate 70.

Just before reaching I-70, the highway crosses the axis of the Salt Valley anticline. Look back to see its collapsed center between ridges of Entrada sandstone.

Utah 9
I-15—Mt. Carmel Junction
51 miles/82 km.

As Utah 9 leaves Interstate 15, it travels eastward across the Virgin anticline, an oval-shaped anticline about 12 miles long but only two or three wide, its center opened up by erosion. There are several similar anticlines in this area. Their bowed-up rocks form natural traps for oil, which collects below impervious rock layers. The Virgin oil field north of the town of Virgin is Utah's oldest, discovered in 1907. Wells here and on adjacent anticlines are relatively shallow — 475 to 800 feet deep. Most of the anticlines in this region have been drilled, but not all contain oil.

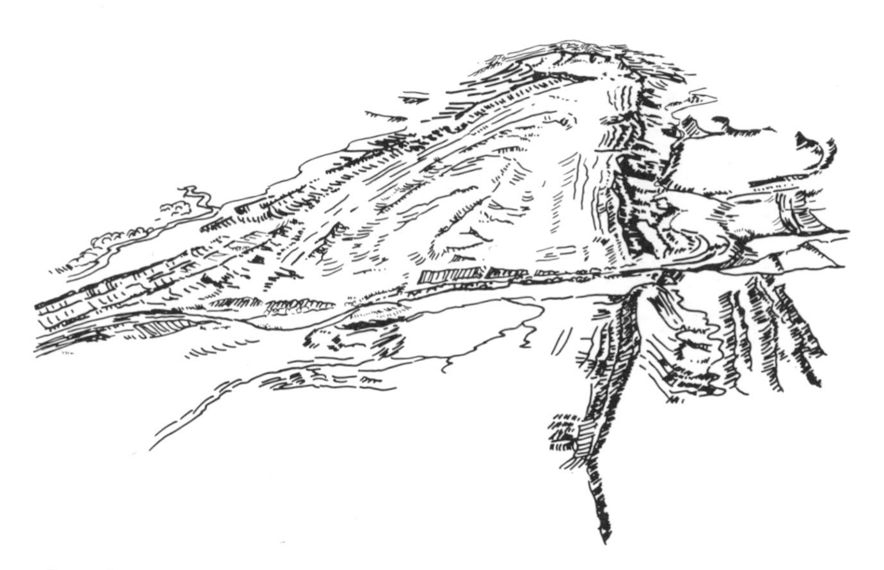

Seen from the air, the north end of the Virgin anticline looks like a giant slipper encircled by ridges or cuestas of tiltedTriassic sedimentary rocks.

Triassic rocks domed up by the anticline are well exposed in roadcuts. The cross-bedded, pebbly Shinarump conglomerate is easy to recognize, as it forms a prominent cuesta that completely rings the anticline. The deep red rock below it is the Moenkopi formation, which contains candy-striped sandstone and gypsum layers. Permian Kaibab

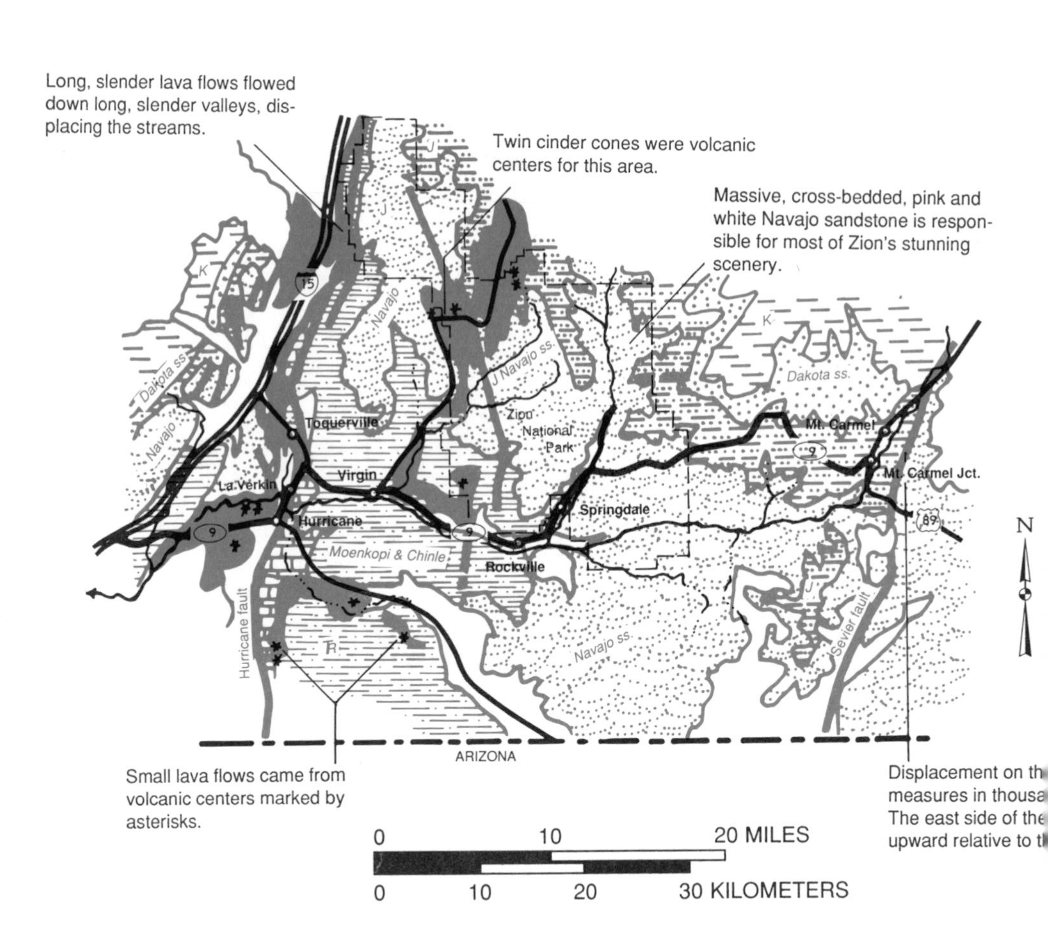

UTAH 9
I-15 — MT CARMEL JUNCTION

Triassic rocks on the flank of the Virgin anticline are faulted (arrow) and crumpled by related earth movements.

limestone cores the anticline, below the soil of the central valley. From milepost 4 it is easy to see how large this anticline is.

Dark basalt lava flows near milepost 5 come from the small volcanic cone to the southeast, visible from milepost 7.

At the bridge across the Virgin River near milepost 11 the river cuts a deep canyon through several lava flows. Some of the flows took the path of least resistance down an earlier channel of the Virgin River, forcing the river to detour to new channels farther west.

The Hurricane fault, a major normal fault that delineates the western edge of the Colorado Plateau, is just east of the town of Hurricane. East of the fault, the same basalt exposed near milepost 5 was raised 200 to 2000 feet. Fault offset at the bridge over the Virgin River between Hurricane and La Verkin (mileposts 10-11) is about 200 feet. Obviously the fault has moved since the lava flows erupted. As is not uncommon along faults, especially where rivers have carved deep canyons, there are hot springs in the gorge of the Virgin River. Their sulfurous steam can often be smelled from the bridge.

The highway crosses the Hurricane fault about a mile east of our highway's junction with Utah 17. Here the country takes on the true character of the Plateau country, a land of intricately sculptured but basically flat-lying sedimentary rocks. As individual plateaus erode, they become mesas, buttes, and then pinnacles; there are plenty of

Zion's massive sandstone cliffs, carved in pink and white Navajo sandstone, are capped with thin layers of the Carmel or Temple Cap formation.
—Ray Strauss photo

examples of this process here. Dark red Triassic sandstone and siltstone north of the highway contrast with tan Permian limestone, mostly sage-covered, south of it. Several small faults complicate the geologic picture. Some of the limestone is highly polished and finely grooved by fault movement; geologists call such polished surfaces slickensides.

The spectacular cliffs of Zion National Park rise to the east. The southern skyline is jagged with lava flows, dikes, and purple cinder cones. Here and there the scene is accented by dark lava flows. To the west the Virgin River narrows as it slices through a lava dam.

The highway proceeds toward Zion National Park across badlands formed in rainbow-colored Triassic rocks — the Chinle formation of Painted Desert fame. Zion's great cliffs and towers become increasingly evident as we continue eastward. Carved in Jurassic Navajo sandstone, prize-winner among scenery-makers in southern Utah, many of the great towers are capped with thin layers of the Temple Cap and Carmel formations. Watch for the columnar jointing in a lava flow north of the highway.

The mountain straight ahead to the east from milepost 24 is Mt. Kinesava. Jurassic limestone and shale, the Carmel and Temple Cap formations, form its cap, with massive Navajo sandstone cliffs below. In the lower part of the cliffs, lacking the large-scale cross-bedding of the Navajo sandstone, are stream deposits of the Kayenta formation, which extends down into the steep slopes below the cliffs. Still lower are the Chinle formation's sculptured badlands. East of Rockville are many more views of these units.

The grandeur of Zion Canyon opens up at Springdale, a town known for its many rock and mineral shops. For a discussion of Zion National Park, see Chapter IV. Except for some Tertiary lava flows, all the rocks there are Triassic and Jurassic.

Our highway climbs out of Zion Canyon by zigzagging up across steep Kayenta formation slopes above Pine Creek, then tunneling through the Navajo sandstone, whose cliffs were, until this highway was built, a barrier to eastward traffic.

Long vertical streaks of carbon mark Zion's stupendous walls. They mark seepage lines where mosses and algae grew on the damp rock. Patches of smooth, shiny desert varnish show up on these cliffs as well.

Sloping layers formed on the leeward sides of Jurassic sand dunes remain today as decorative cross-bedding, here accentuated by erosion. Horizontal bands between sets of cross-bedding commonly contain silt and sand deposited in flat areas between the dunes.

East of the tunnel are many unusual exposures of Navajo sandstone. Its decorative, large-scale cross-bedding developed as successive layers of fine dune sand, blown up gentle windward dune slopes, were deposited on steeper leeward slopes, where wind velocity lessened. Numerous vertical joints in this rock became pathways for movement of groundwater. The groundwater in turn weakened the rock by dissolving some of the calcium carbonate that cemented its grains together. Helped by plant roots seeking moisture, many joints widened into narrow defiles and "secret" canyons.

Outside the park the highway crosses a ravine-creased landscape still surfaced with Navajo sandstone, its ornate cross-bedding showing up plainly and in many forms. Some of the little valleys on this surface don't drain at all. Rainfall and snowmelt just sink in, percolating downward through the Navajo sandstone, ultimately feeding springs in Zion and other deep canyons. Watch for the recent gullying in upland meadows here.

Continuing eastward, we gradually see more and more of the rock layers above the Navajo sandstone. A little west of milepost 54, for instance, the Carmel formation, cut by small faults, appears near the highway. Farther east, greenish brown soils derive from the Cretaceous Tropic shale. This poorly consolidated formation makes a poor roadbase, hence the rough road, with many patches, and evidence of slumps between mileposts 49 and 50. In all of the Plateau country, Cretaceous rock layers are brownish or grayish, whereas Triassic and Jurassic strata tend to be pink and red, a difference caused by differences in oxidation of iron minerals.

The White Cliffs of Utah's Grand Staircase are now in view to the east and southeast. On them you'll recognize the Navajo sandstone, carved with the Carmel and Entrada formations. It's easy to see that the Navajo sandstone is higher on these cliffs than it is in canyons below and south of our highway. It has been lifted along the Sevier fault, which separates the Markagunt Plateau from the Paunsaugunt Plateau to the east. Displacement along this fault amounts to several thousand feet.

The highway descends toward Mt. Carmel Junction through soft Cretaceous rocks, thickly soil-covered, on the downthrown side of the Sevier fault. In a roadcut near the junction are some interesting stream gravels, evidence that streams in this area have at some time shifted their courses. The present valley of the East Fork of the Virgin River lies right along the Sevier fault.

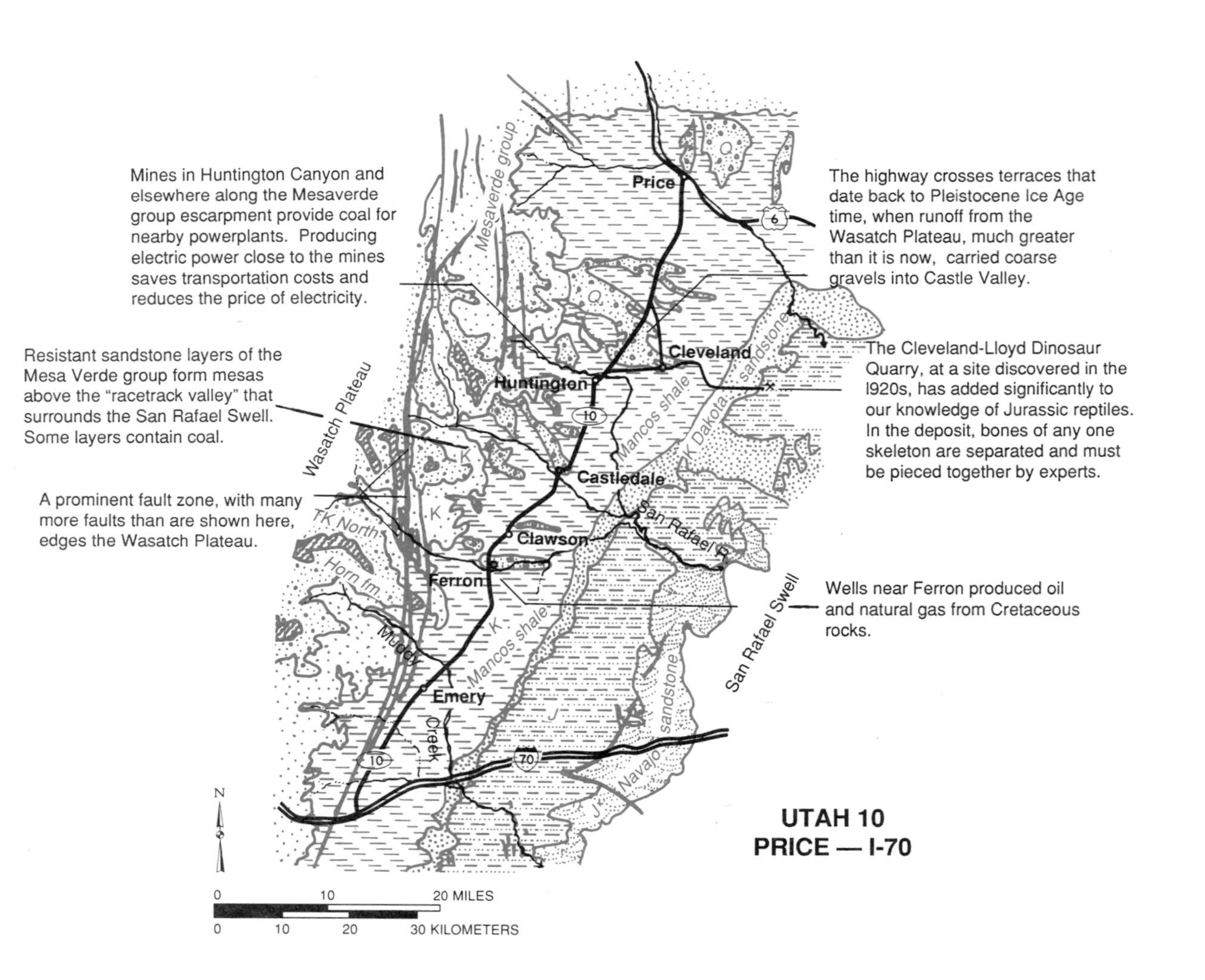

UTAH 10
PRICE — I-70

Utah 10
Price — I-70
69 miles/111 km.

West of Price, the Book Cliffs curve southward, hiding much of the higher Wasatch Plateau. The cliffs are capped with Cretaceous Mesaverde sandstone. The town of Price lies on river-formed gravel deposits in a broad valley eroded into soft Mancos shale. Long tongues of coarse gravel and sand extend from the cliffs, forming mesas around the town; they reach well to the south along our highway.

A few miles south of town, the highway climbs onto one of these terraces. From this higher vantage point you can see the low dome of the San Rafael Swell farther south — a major topographic feature in which Paleozoic and Mesozoic rocks arch upward in an 80-mile-long, 40-mile-wide anticline. The anticline is lopsided: Its western slope, which we'll be seeing on this route, is more gently inclined than its eastern slope.

Section across Utah 10 near Cleveland

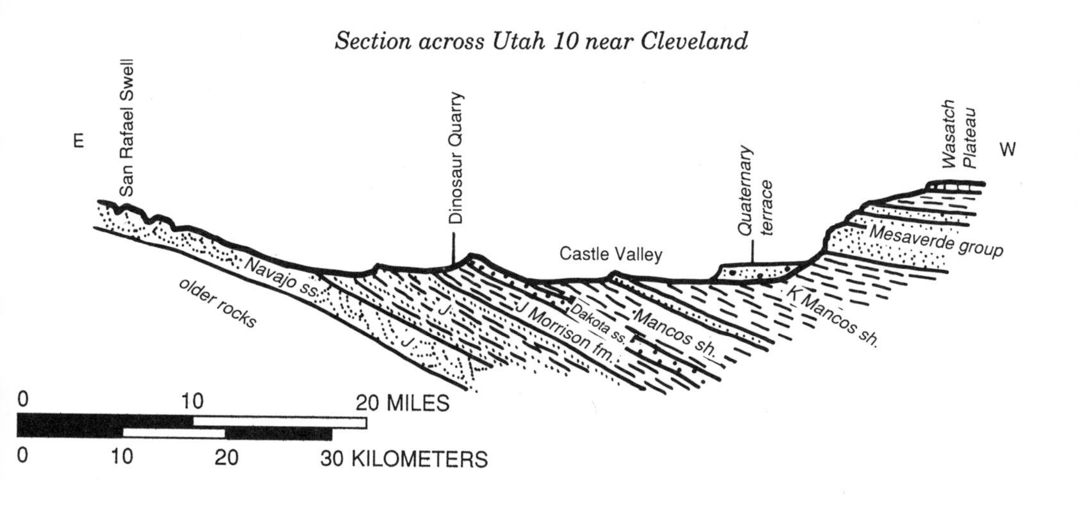

A skeleton of Allosaurus fragilis, *a carnivorous dinosaur, is exhibited in the Cleveland-Lloyd Dinosaur Quarry visitor center.* —BLM photo.

South of milepost 57, a road to the left goes to the Cleveland-Lloyd dinosaur quarry. There, paleontologists from Brigham Young University, the University of Utah, Carnegie Museum, and Princeton University have unearthed skeletons of several types of Jurassic dinosaurs. They were found in the Morrison formation, as were those of Dinosaur National Monument in northeastern Utah. The ponderous reptiles seem to have become mired in fine, clayey muds of a prehistoric bog or river floodplain. Both carnivorous and herbivorous varieties are represented, among them *Camarasaurus, Ceratosaurus, Allosaurus, Stegosaurus*, and *Camptosaurus*. Dating of volcanic ash that overlies the dinosaur-bearing sediments indicates that they are about 147 million years old. The quarry is now a National Natural Landmark.

The highway crosses one terrace after another, dropping between them to the Mancos shale of the valley floor, part of a racetrack valley around the San Rafael Swell. Wherever the highway crosses Mancos shale, the pavement is bumpy.

As the highway continues south along Castle Valley, between the San Rafael Swell and the Wasatch Plateau, it passes through part of the San Rafael oil field, no longer productive. Visible along the highway are various small domes with rock layers arching over them —just the kind of domes that, when they occur underground, act as traps for oil and gas. There is a dome, for instance, to the left between mileposts 41 and 40. The oil is in cavities and pore spaces in Permian limestone.

The big Hunter Powerplant south of Castle Dale (as well as a less conspicuous plant northwest of Huntington) obtains coal from mines in nearby Cretaceous rocks. Both the Mancos shale and the Mesaverde formations contain coal. The plants burn low-sulfur, low-ash coal and employ pollution-control technology as well. Power from both plants goes into a grid that serves parts of Utah, Wyoming, and Idaho.

Castle Dale gets its name from the castellated turrets on the sides of some of the Quaternary terraces edging the valley. Formed of fine sediments containing volcanic ash, many of these terraces are no longer connected with the mountains they once flanked, but stand as isolated, steep-sided mesas.

Where the Book Cliffs approach the highway, a roadcut exposes the contact between dark gray Mancos shale and overlying Mesaverde siltstone, sandstone, and thin coal seams. In the background the contact falls at the base of the cliffs, and the Mancos shale is concealed beneath fallen rock of the talus slope.

As soft shales of the Mancos formation weather and erode, boulders of overlying lava tumble down across them.

South of Ferron the highway, still going up and down successive terraces, converges with a fault zone that edges the Wasatch Plateau and separates the Plateau country discussed in this chapter from the High Country of Chapter II. The faults disappear southward under the volcanic mountains of the Fish Lake High Plateau, visible to the south near I-70. These mountains are part of Utah's largest volcanic area, where vast quantities of Tertiary and Quaternary lava, volcanic ash, and breccia cover a region 70 miles across. Dark lava boulders washed from this region can be seen near the highway between milepost 3 and Interstate 70.

Utah 24
Capitol Reef National Park — I-70
84 miles/135 km.

East of the Capitol Reef National Park visitor center, Utah 24 follows the Fremont River through its narrow canyon, penetrating Capitol Reef, a high ridge of steeply tilted Mesozoic strata. The highway passes in sequence a succession of these rocks as they bend downward along a prominent monocline known as Waterpocket Fold — the feature attraction of the national park. For a geologic section through Waterpocket Fold, see Capitol Reef National Park in Chapter IV.

From the dark red Moenkopi formation near the visitor center to the gray Mancos shale east of the monocline, the highway passes through younger and younger rocks. Just above the Moenkopi formation are gray-green shales of the Triassic Chinle formation, and above them, forming perpendicular cliffs dark with desert varnish, the Wingate sandstone, now thought to be Jurassic in age.

In rocks near the highway, numerous small, deep, rounded hollows mark otherwise barren rock surfaces. Such hollows, called tafoni, are especially common in the Wingate sandstone and in the massive, light-colored Navajo sandstone, the rock that makes up the eastern face of Capitol Reef and the white domes that give it its name. What

Wide-sweeping cross-bedding etched by wind and rain characterizes the Navajo sandstone. Small cavities may deepen to form the "waterpockets" for which Waterpocket Fold is named.

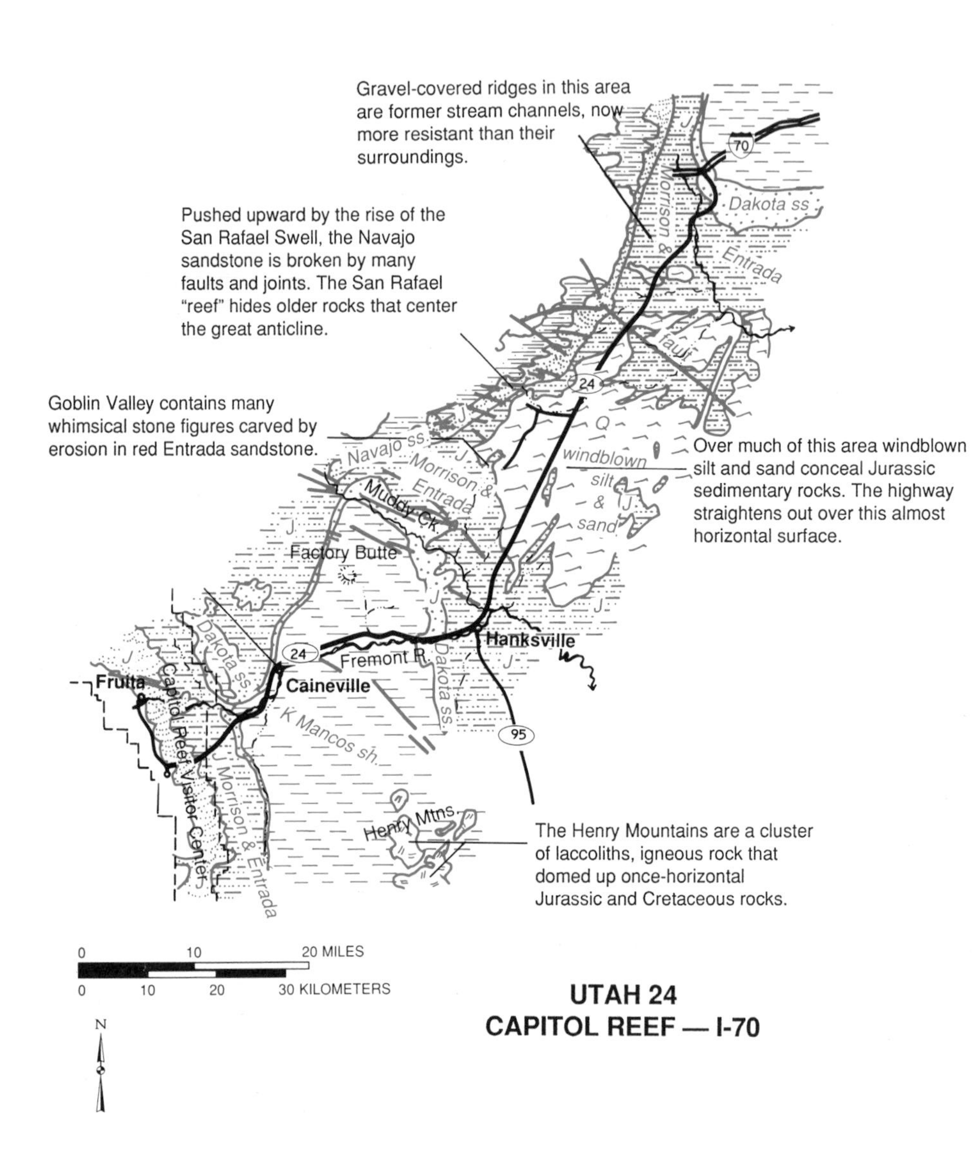

UTAH 24
CAPITOL REEF — I-70

Overloaded with rock debris, the Fremont River blocks its own shallow channels with gravel and sand. Thus it must frequently shift its course, creating a braided channel.

initiates these little hollows is unknown; they may begin with loosening of sand grains around salt crystals or other soluble minerals in the rock. Since most of the holes are high above river level we know that it is wind and not water that whirls and ultimately removes loosened grains, hollowing out the cavities. Larger cavities within the Navajo sandstone have widened into the "waterpockets" that give the fold its name.

The Fremont River heads in Fish Lake on the Fish Lake Plateau. There it flows northeastward, curving around gradually into a southward and then southeastward course. East of milepost 86 it formerly swung in a tight loop around an isolated rock mass. When this highway was built, the river was provided with a new channel and a manmade waterfall. Big gray basalt boulders along the river were brought here by a more powerful ice-age stream fed by glaciers in high country to the west.

Emerging from the canyon between mileposts 91 and 92, the highway enters Utah's Painted Desert, with outcrops of Jurassic Morrison formation on both sides of the highway. This formation contains an abundance of clay that swells and shrinks as it wets and drys. Few plants manage to grow on its loose surface. The sloping cuesta east of the barren badlands is capped with tan Dakota sandstone, the lowest, oldest Cretaceous rock in the Plateau country. Still farther east are drab gray hills of Mancos shale, some of them wearing thin, flat caps of sandstone.

The Mancos shale cliff north of milepost 102 sparkles with crystals of gypsum. Some outcrops contain dark coaly layers; others contain white bentonite zones. Factory Butte to the north is capped with a hard sandstone layer, as are the smokestacks of Steamboat Point to the south. The rocks have leveled out here east of Waterpocket Fold.

Fossil oysters are common near mileposts 111 and 112, in the lowest layers of the Mancos shale and in the Dakota sandstone below it, as well as in tumbled blocks near the road. Terrace gravels north of here are composed almost totally of redeposited fossil oyster shells; in places they have even been quarried for road material!

The route has crossed a syncline and is getting into older rocks again. Near Hanksville, roadcuts again display colorful shales of the Morrison formation. Castled brown bluffs near Hanksville are in the Summerville formation, deposited on river floodplains, deltas, and lake flats of Jurassic time. Watch for Jurassic stream channels in these rocks.

Hanksville lies near the confluence of the Fremont River and Muddy Creek. The two streams come together to form the Dirty Devil River, named by John Wesley Powell during his exploration of the Colorado River in 1869. Where it flowed into the Colorado, Powell and his men found the water "exceedingly muddy and (with) an unpleasant odor," and when one of the men said it was a dirty devil, the name stuck.

Roadcuts north of the bridge between mileposts 119 and 120 expose the Jurassic Entrada formation. Depending on its grain size — sand or silt or mud — and on the tightness of the limy or siliceous cement that holds the grains together, this formation erodes into arches, as in Arches National Park, into cliffs, as in buttes near this highway, or into fanciful little hoodoos, as at Goblin Valley State Park just north of here.

At the viewpoint south of milepost 124, pointers identify topographic features of this area. Many are remnants of Entrada sandstone, and show that weak, crumpled siltstone near the base of the formation is responsible for undermining many sandstone cliffs. Lava-capped Thousand Lake Mountain and Boulder Mountain show up to the west and southwest. Goblin Valley is to the west — you can just make out some of the stone goblins.

Gilson Butte and Little Gilson Butte mark the skyline to the south. They, too, are decoratively carved remnants of the Entrada formation. The Henry Mountains farther south are a cluster of igneous intrusions that in Oligocene time pushed up through Paleozoic, Triassic, and

In Goblin Valley the Entrada sandstone has eroded into hundreds of weird and amusing figures. Erosion of weak siltstone and mudstone layers undermines stronger sandstone, producing the unusual rock forms.

Jurassic sedimentary rocks, doming up overlying Cretaceous formations. Cretaceous rocks are now eroded from their summit, but occur in upturned rings around the base of the mountain; remnants of them can also be found between the individual intrusions.

Goblin Valley State Park is worth a visit, if only for fun and photographs. The quaint goblins, along with seals, elephants, and other bizarre figures, are given form and symmetry by alternating hard and soft layers of mudstone, sandstone, and siltstone. White rock above the brick red Entrada formation on nearby cliffs is the Curtis formation, with the Summerville and Morrison formations above it.

The Goblin Valley road goes near the mouth of Chute Canyon, where the Navajo sandstone forms a high hogback along the face of the San Rafael Swell. The hogback extends northeastward as San Rafael Reef, which will be in sight the rest of the way to the Interstate 70 junction. The white rock is deeply scored by canyons that formed along its numerous vertical joints.

Much of the vast and uninhabited San Rafael Swell didn't really open up until the uranium boom of the 1950s and '60s. Then, when Jurassic rocks of this area were found to contain uranium, many mines were developed. All are now inactive. The area also contains oil-impregnated rock, not mined at present but remaining as a reserve until such time as rising oil prices make mining and processing profitable. Occurring in the Permian White Rim sandstone, the oil is in the form of elaterite, a tarry brown substance that seeps from the rock in a number of places.

Continuing northeastward across the gravel-veneered pediment below the San Rafael Swell, we get numerous good views of the great anticline and of the sedimentary rocks that cover it. Small dunes edge the highway in places, their sand derived mostly from the Navajo

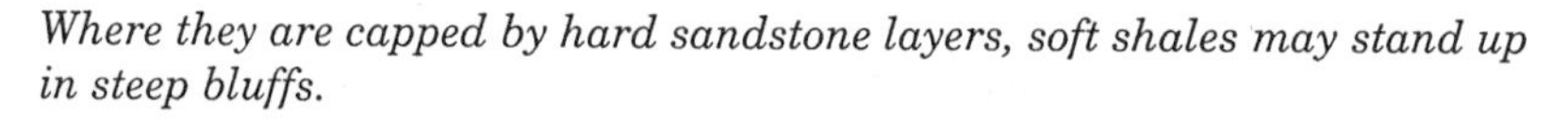
Where they are capped by hard sandstone layers, soft shales may stand up in steep bluffs.

Near the I-70 junction, Morrison formation shales lend color to barren, eroded slopes.

sandstone of the San Rafael Reef. Some of the sinuous, gravel-covered ridges between the reef and this stretch of highway were once stream channels. As the gravel was impregnated and cemented with caliche, a deposit of calcium carbonate and other minerals that forms as alkaline groundwater evaporates, streams eroded new channels into soft, unprotected former ridges. So the topography is reversed: ridges where there once were streams, and channels where there once were ridges. Gravel-filled channels show up in roadcuts as our highway approaches Interstate 70.

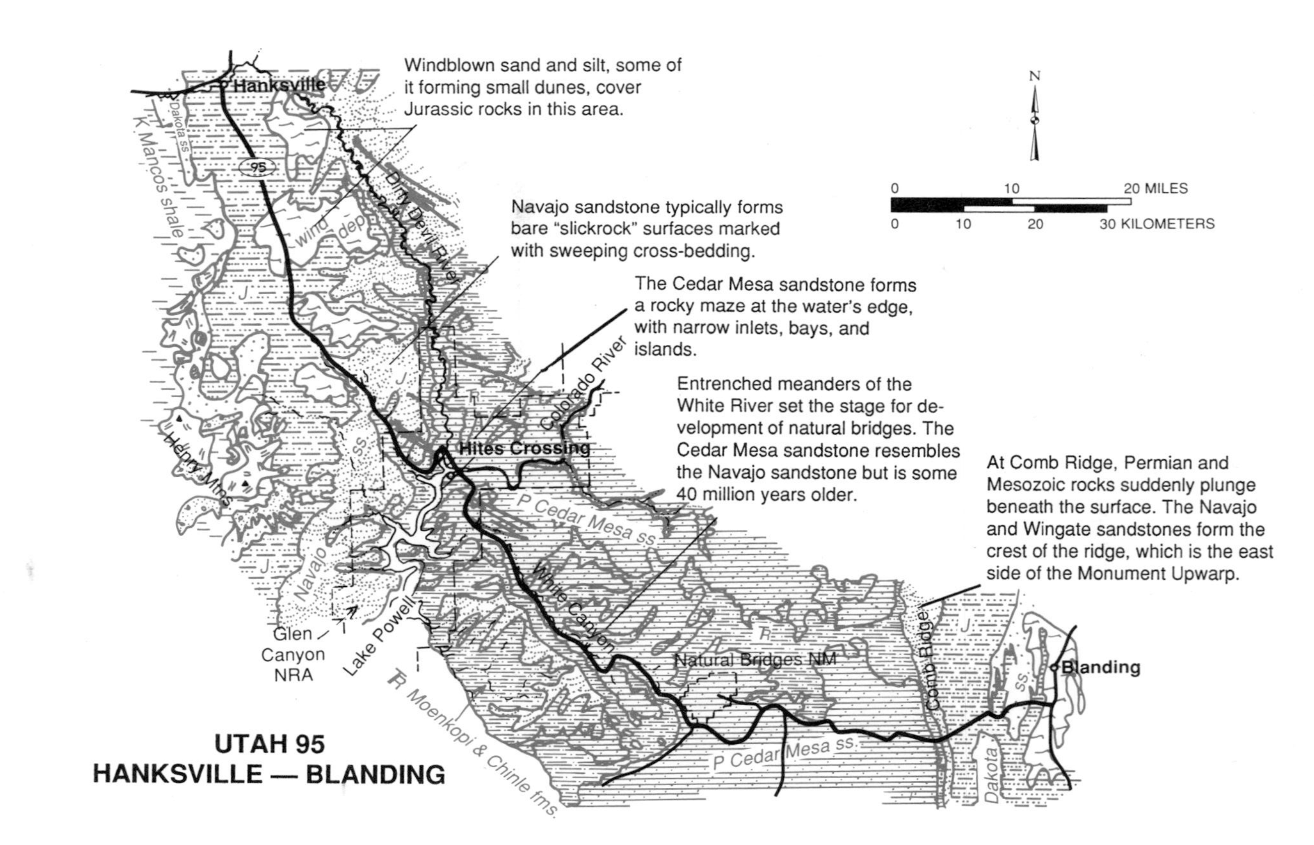

UTAH 95
HANKSVILLE — BLANDING
Windblown sand and silt, some of it forming small dunes, cover Jurassic rocks in this area.
Navajo sandstone typically forms bare "slickrock" surfaces marked with sweeping cross-bedding.
The Cedar Mesa sandstone forms a rocky maze at the water's edge, with narrow inlets, bays, and islands.
Entrenched meanders of the White River set the stage for development of natural bridges. The Cedar Mesa sandstone resembles the Navajo sandstone but is some 40 million years older.
At Comb Ridge, Permian and Mesozoic rocks suddenly plunge beneath the surface. The Navajo and Wingate sandstones form the crest of the ridge, which is the east side of the Monument Upwarp.
N
0
10
20 MILES
0
10
20
30 KILOMETERS
Hanksville
95
Dakota ss
K Mancos shale
Dirty Devil River
wind dep
J
Henry Mtns
Navajo
ss
Glen Canyon NRA
Lake Powell
Hites Crossing
Colorado River
P Cedar Mesa ss
White Canyon
TR Moenkopi & Chinle fms.
Natural Bridges NM
P Cedar Mesa ss
Comb Ridge
Dakota
Blanding

Utah 95
Hanksville — Blanding
132 miles/212 km.

Utah 95 leaves Hanksville on a surface of wind-blown silt and sand, with small dome-shaped dunes held in place by clusters of plants. Along the road are occasional outcrops and "goblins" or "stone babies" of dark red mudstone in the Entrada formation.

Jurassic rocks surface a large area here. Above the Entrada formation are the Curtis, Summerville, and Morrison formations — all fairly weak and weathering back in wide benches that alternate with ledges of harder sandstone. A few sinuous, ridgelike mesas, including those near mileposts 2, 3, and 4, are protected by caps of old stream gravels, some of which are partly cemented with white caliche, which forms in arid regions.

The gravels contain pebbles of gray igneous rock from the Henry Mountains to the south. This free-standing mountain range consists of a cluster of Tertiary intrusions that pushed upward through Paleozoic, Triassic, and Jurassic rock layers, but domed up overlying Cretaceous strata. The igneous rock is diorite porphyry, diorite meaning it has more of certain dark-colored minerals than granite, porphyry meaning that it includes large crystals embedded in a finer groundmass. Though most of the domed Cretaceous rocks have since

Overlying Cretaceous sedimentary rocks were tilted and lifted as molten rock of the Henry Mountains pushed upward. Today they ring its base and appear in low saddles between the summits.

eroded away, remnants of them still fill saddles betwen the individual intrusions. There is no obvious sign of glaciation on these mountains; horseshoe-shaped scarps that look very like glacial cirques are due to landslides.

Both Triassic and Jurassic sedimentary rocks in this area contain uranium, and this region was thoroughly prospected during the uranium boom of the 1950s and '60s. The sediments were deposited in lake, marsh, and delta environments, and contain abundant organic matter — plant and animal material. Derived from volcanic rock, uranium minerals were dissolved and transported by groundwater moving slowly through the rocks over millions of years. Wherever there was organic matter, the uranium minerals came out of solution and settled in the rock. Most of the uranium mines in this area developed where there were sizeable concentrations of organic matter and therefore of uranium.

Watch for small folds in Jurassic rocks near the Garfield County line. There are more Entrada formation goblins here as well. Farther south, the highway goes up and down through the Jurassic formations, which are well exposed in their proper sequence west of mileposts 21 and 22: the goblin-forming Entrada formation at the base, and light-colored Curtis, Summerville, and Morrison formations in sequence above it.

Just south of milepost 23, the highway encounters some pink, cross-bedded Navajo sandstone that accumulated in dunes on a Jurassic desert that stretched from Nevada to Wyoming. Typically the Navajo sandstone forms barren slickrock surfaces that vary in

Pebbles and cobbles scattered on a Navajo sandstone surface are diorite porphyry, intrusive igneous rock from the Henry Mountains. Dark rock near the large shrub is red siltstone of the Kayenta formation. Two small faults offset the Navajo-Kayenta contact.

The Wingate sandstone surface at the left is shiny with desert varnish and deeply pitted by wind and rain. To the right, where slabs of rock have broken away, desert varnish has not yet developed, and the cross-bedding of this rock can be seen. At far right, streaks of lichens are cut off abruptly by the most recent rockfall.

color from white to coral pink. Coral-colored sand dunes near the junction with Utah 276 are derived from this rock — new dunes from old.

In places it is difficult to tell the Navajo sandstone from the Wingate sandstone below it, which first appears near the Utah 276 junction. The Wingate, also a Jurassic dune deposit, tends to form sharp-angled, pitted cliffs darkened with desert varnish. In both the Wingate and the Navajo sandstones the large-scale, dune-style cross-bedding is here and there interrupted by discontinuous horizontal bands of fine silt and clay. These bands are interdune deposits that represent swales between individual dunes, places where windblown silt and clay accumulated. The two formations are separated by horizontal beds of red siltstone and sandstone of the Kayenta formation.

As the highway descends through the canyon of North Wash, these formations show up clearly, and their similarities and differences can be seen. A nature trail at the rest stop south of milepost 33 gives even better opportunities to examine the Mesozoic rocks. Below the Wingate formation are colorful bluish and purplish mudstone of the Chinle formation and, still lower, dark red-brown marine siltstone and

Where a block of Chinle mudstone has slid downward along a curving surface, the once flat-lying beds are tilted backward—toward the right. Normally the Chinle is a slope-former; here downcutting has been so rapid that the formation is well exposed.

mudstone of the Moenkopi formation, both Triassic in age. Deposited on a wide floodplain dotted with shallow ponds and marshes, the Chinle formation contains lots of volcanic ash. The ash has altered to clay that swells with every rain, thus discouraging vegetation. This rock unit also contains a good deal of mudstone and several sand and conglomerate layers, the lowest of which forms a hard, angular ledge at the base of the formation, right at the Glen Canyon National Recreational Area sign.

The Moenkopi formation's red mudstone, siltstone, and sandstone are products of an ancient bay at the edge of a western sea, a wide, shallow bay where mud, silt, and sand, washed from a highland near the Colorado-Utah border, accumulated in well-ordered layers. From the Hite overlook, dark red or chocolate-colored rocks of this formation descend all the way to Lake Powell. Farther upstream, older rocks edge the lake: the Cedar Mesa sandstone of Permian age, a unit that is partly near-shore sandstone, partly dune sand. This resistant unit is carved into enticing narrow side canyons and, along the lake, innumerable small bays. It weathers into rounded surfaces pocked with wind-formed hollows. The cross-beds in the Cedar Mesa sandstone formed under the easterly winds of the tropics, and dipped west until they were rotated counterclockwise as the continent moved away from Europe.

Here at its upper end, Lake Powell's water is murky. Sediments carried by the Colorado River and its tributaries have not yet settled out. Coarse sediments are gradually building a delta into this part of the lake; finer sediments travel farther downstream to settle into the deep waters of what was once Cataract Canyon. Before Glen Canyon Dam was built, Cataract Canyon held a white-water stretch of the Colorado River that challenged river-runners from the time of John Wesley Powell's first trip in 1869. Downstream, beautiful Glen Canyon, where the river coursed gently among towering cliffs of Navajo and Wingate sandstone, was drowned as well.

Watch for ancient ripplemarks on siltstone slabs as the highway descends through the Moenkopi formation. There are good opportunities to see the Cedar Mesa sandstone as the highway approaches the bridges that span the Dirty Devil and Colorado rivers.

Southeast of the two bridges the strata rise eastward onto the Monument Upwarp, with several faults offsetting the contact between the Cedar Mesa sandstone and the Moenkopi formation. Near milepost 52 are good views west to the entire sequence of rocks described above, against a backdrop of Mt. Holmes and Mt. Elsworth, the southernmost of the Henry Mountains intrusions.

Small side roads in this area lead to uranium mines and prospects, some of which can be spotted from the road. The uranium concentrates within stream channels in the Chinle formation, where dead and dying plant and animal material collected in Triassic time.

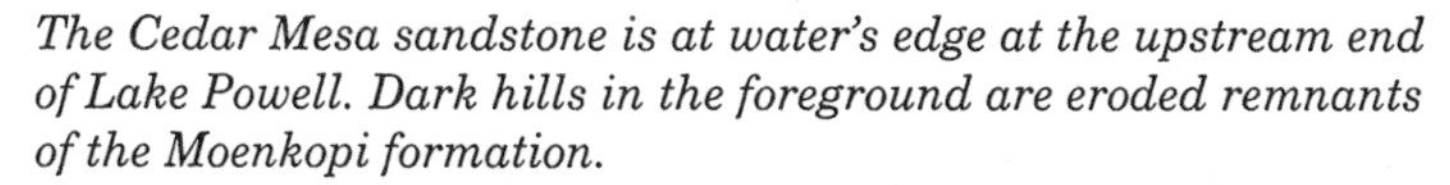

The Cedar Mesa sandstone is at water's edge at the upstream end of Lake Powell. Dark hills in the foreground are eroded remnants of the Moenkopi formation.

Comb Ridge, with the Navajo sandstone just below its crest, extends southward and southwestward to Kayenta, Arizona.

Near the highway, the White River swings in looping meanders inherited from a time before canyon-cutting began, when it wound lazily across a nearly flat surface. With regional uplift and possibly a wetter climate to increase its cutting power, the river sliced down through the Cedar Mesa sandstone; it now flows some 60 feet below the general surface. Such entrenched meanders often set the stage for natural bridges like those at Natural Bridges National Monument, described in Chapter IV.

Strata continue to rise toward the crest of the Monument Upwarp between mileposts 94 and 97. The view to the south shows that the low dome of this immense anticline, which formed at the end of the Mesozoic Era, extends well into Arizona. East of the crest we begin a long descent across the sloping upper surface of the Cedar Mesa sandstone. In the distance ahead, in Colorado, is the Sleeping Ute, another cluster of diorite intrusions similar to the Henry Mountains. The north end of the Carrizo Mountains in Arizona show up to the south. The Abajo Mountains, yet another cluster of intrusions, are visible to the north from mileposts 102 and 104.

The slope steepens east of milepost 104, plunging into the valley of Comb Wash. There and on Comb Ridge across the valley are the same Triassic and Jurassic strata we saw between Hanksville and Lake

Powell: the Moenkopi and Chinle formations near Comb Wash, and the Wingate sandstone, Kayenta formation, and Navajo sandstone on Comb Ridge. All dip steeply eastward as the east flank of the Monument Upwarp.

Near milepost 118 the road climbs onto a tableland surfaced with Dakota sandstone, the oldest Cretaceous unit in this region. Weak purple and green mudstone of the Morrison formation appears in canyons that cut this surface. Near the junction with US 191 the rocks lie hidden beneath a veneer of windblown silt and sand.

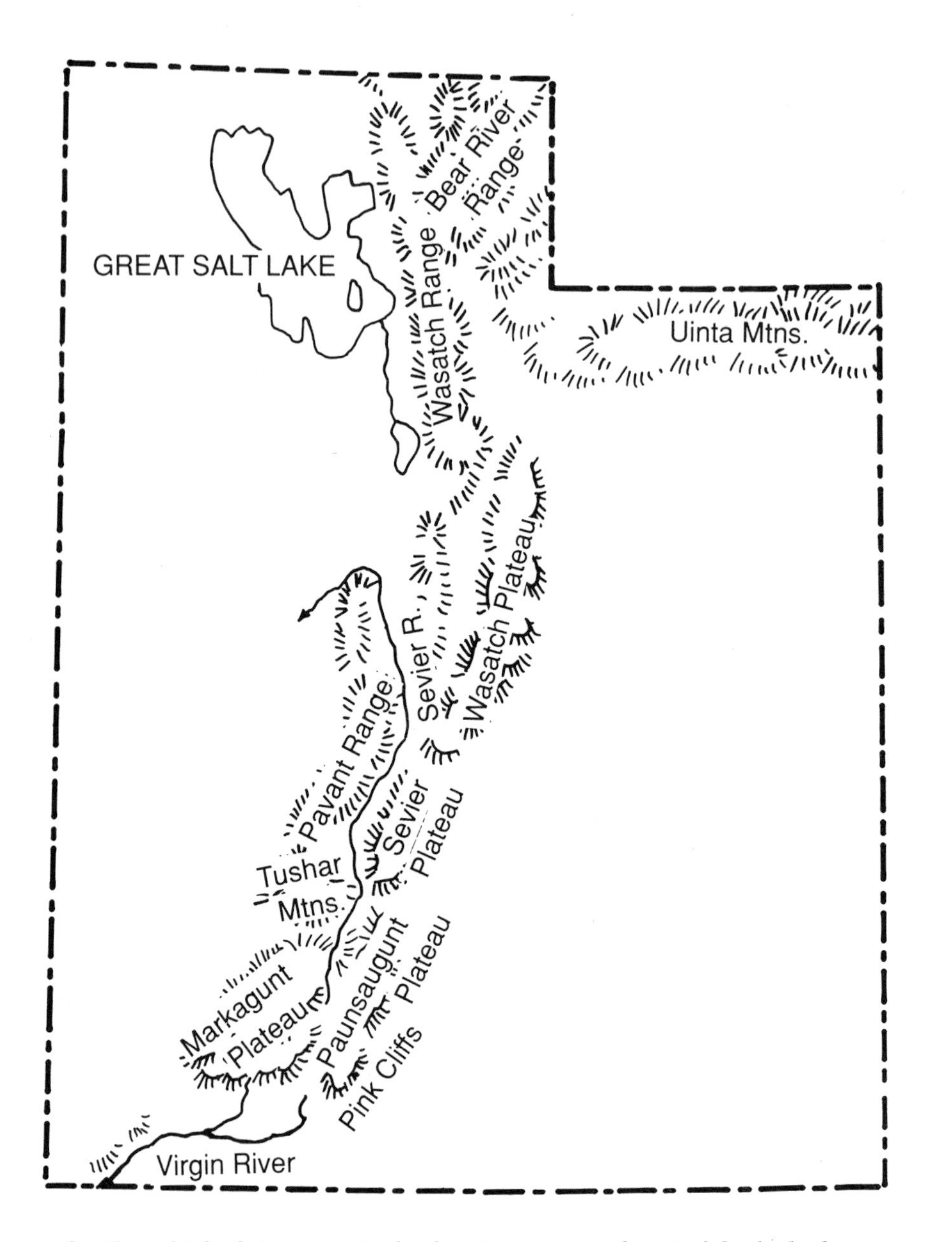

Southern high plateaus, central volcanic ranges, and part of the faulted, folded Southern Rocky Mountains in the north make up Utah's High Country. The Sevier Valley, following a line of a north-south fault, divides plateaus and ranges.

II
Utah's Backbone
The High Country

Utah's highest mountains, running like a backbone down the center of the state, come in many shapes and varieties. Geologically, they fall into three basic patterns:

The Uinta Mountains in the northeast corner of the state are one big anticline faulted down its center and along both its edges. An east-west structure in a state otherwise full of north-south geologic patterns, the range extends westward almost to the Wasatch Range; its anticlinal structure extends right through that range, telling us that the anticline existed before the faulting that raised the Wasatch Range. Technically, the Uinta Mountains are part of the Rocky Mountains.

The Wasatch Range is more complex, a product of several superimposed episodes of faulting and folding. The steep western face of the range, the Wasatch Mountain Front, is the line of offset on the Wasatch fault, the easternmost major normal fault of the Basin and Range province. This zone of normal faulting began movement at the same time as faults in the Basin and Range area to the west. The flat-faced triangles of truncated mountain spurs, as well as small fault scarps crossing alluvial fans, show that some of the movement on the Wasatch fault is geologically quite recent. Frequent minor earthquakes with epicenters along the Wasatch fault warn us that movement continues to this day.

High country south of the Wasatch Range bears the simplicity of the Plateau country: basically horizontal layers of sedimentary rock, broken along faults and raised differing amounts into the so-called High Plateaus. These tablelands, demarcated on the west by the Hurricane and other faults, are covered over much of their extent by volcanic rocks, including great thicknesses of volcanic ash or tuff, and hard, resistant lava flows that help them to maintain their present high elevations.

Rocks exposed in these three regions range in age from Precambrian gneiss, schist, and partly metamorphosed sedimentary rocks at the core of the Wasatch and Uinta mountains, to Tertiary lake deposits and Tertiary and Quaternary volcanic rocks of the High Plateaus.

The oldest of the Precambrian rocks lie along the steep Wasatch Front north of Salt Lake City. They consist of hard, highly altered metamorphic rocks, schist and gneiss about 2.6 billion years old. Younger Precambrian sedimentary rocks, roughly a billion years old, core the Uinta Mountains and show up also in the Wasatch Range south of Salt Lake City, where the Uinta Mountain anticline extends west to the Wasatch fault. These younger Precambrian rocks are thousands of feet thick. Their sedimentary origins can easily be recognized in marble that was once limestone, slate that was once shale or mudstone, and quartzite that was once sandstone. They even include some tillite, a kind of rock derived from bouldery gravel deposited by glaciers.

Paleozoic sedimentary rocks surround these Precambrian areas. Almost all are marine, deposited in a shallow sea that spread unhindered across the gentle western slope of the continent, in a picture similar to our modern east coast. Cambrian sandstone at the bottom of the Paleozoic sequence is a near-shore sandstone; it gets younger eastward, telling a clear story of the sea invading the continent from the west. From time to time during the Paleozoic Era, localized basins deepened and collected great thicknesses of sediments. At other times the land warped upward, forcing the sea to back away to the west.

An interesting feature shown by these rocks in Utah is an apparent north-south hingeline that divided areas of shallow sea from those of deepening sea farther west. The hingeline falls along the western edge of the High Country, roughly paralleling Interstate 15 between Salt Lake City and St. George. West of it, Paleozoic sedimentary rocks thicken rapidly, reaching in some cases thousands of vertical feet.

In parts of Utah's High Country, some of the Paleozoic sequence is duplicated by several large thrust faults that in Cretaceous time slid sheets of sedimentary rock — thick ones from west of the hingeline — eastward from the Sevier Uplift, a mountainous area that developed in western Utah and eastern Nevada in Cretaceous time. Though the Sevier Uplift was west of Utah's present High Country, it left its mark on the High Country in the form of these faults and the coarse conglomerate of alluvial fans that once surrounded the uplift.

Early Mesozoic rocks in the northern part of Utah's High Country are mostly eroded away or covered by rocks thrust eastward as part of the Sevier thrust belt. They do appear along the flanks of the Uinta Mountains, in the Wasatch Range east of Salt Lake City, and, in abundance, on the High Plateaus, especially in areas that are not covered with volcanic rocks. Most Mesozoic strata in these areas are continental, deposited on land, though those that accumulated toward the end of the era record a last invasion of the sea — a sea that this time came from the east.

Four major geologic events made the Cenozoic era memorable, and led to the mountainous terrain and high plateaus we see in Utah today:

Starting late in Cretaceous time and continuing into early Tertiary time, the westward drift of North America apparently set up stresses and strains that broke and buckled the Earth's crust, forming the Rocky Mountains, of which Utah's Uinta Mountains are a part.

The faulting and folding opened up fissures and cracks through which molten magma erupted. Most of the lava and volcanic ash produced in mid-Tertiary time, 40-20 million years ago, was silicic — thick, sticky, fairly light in color, derived in part at least from remelted continental rocks. Such magmas tend to build up tall volcanic cones, and repeatedly cork their own vents with plugs of hardened lava. From time to

Tertiary volcanic rocks include thick lava flows, layers of volcanic ash, and breccia or broken lava and ash.

time, these volcanoes exploded, filling the air with volcanic ash and initiating mudflows that carried ash and broken volcanic rocks down the mountain flanks.

In Tertiary time, as recently as 35 or 45 million years ago and continuing to the present, Utah and adjacent states were lifted upward 3000 to 5000 feet. Accompanying this broad doming, block faulting caused by crustal tension raised some regions, dropped others, creating the Wasatch Range and the many smaller ranges of western Utah. Volcanism associated with this uplift produced basaltic lava that ran down stream valleys or pooled on flat surfaces and behind ridges of other rock, emphasizing and in many cases preserving the plateau quality of the landscape. Basalt lava flows rise from a deeper source than do silicic lavas — from deep under the crust, in the Earth's mantle. They flow freely, without explosions, or if they have absorbed some groundwater and turned it rapidly to steam, puff out bits of bubble-filled scoria that form cinder cones.

Uplift encouraged erosion, which then as now worked to whittle down the mountains and to carve deep canyons. In Pleistocene time, changes in climate brought glaciers — potent agents of erosion — to high mountain valleys, and added to the effectiveness of streams and rivers. Another ice age product was the wide expanse of Lake Bonneville west of the mountains. This great freshwater lake in its turn etched its shorelines on the mountain front.

Utah's high mountain country is generally steep. Steepness — particularly in weak Mesozoic and Cenozoic rocks, is unstable. Helped along by heavy winter snows and summer rains that both load and lubricate the rocks, the mountains see frequent landslides and, in winter, snowslides or avalanches. Some of the slides dam rivers and streams. Where the loose rock of natural dams later breaks through, downstream floods occur. Particularly in the Wasatch Range, many slides — old and new — can be seen from the highways described in this book, some of them initiated by highway construction. The slides serve to remind us that geologic processes go on today as they did in the past.

A small slide near Utah 12 displays two landslide elements: an arcuate scarp at the top, and catsteps on the slide itself. Such slides damage highways by pushing the pavement up from below.

Mines and quarries dot Utah's mountainous backbone, providing the state with iron, manganese, tungsten, uranium, gypsum, and other minerals, as well as with coal and limestone.

Many streams flowing through these mountains are now dammed for their hydroelectric potential and for domestic and industrial water. Heavily populated urban centers along the mountains' western front require more water than the relatively abrupt western slope of the mountains can provide, so complex networks of tunnels and aquaducts now bring eastern-slope water through the Wasatch Range to west-side streams and reservoirs.

Colorful lichens ornament white Precambrian quartzite in the Uinta Mountains. Lichens are pioneers of the plant world, the first to colonize new rock surfaces.

Interstate 70
Fremont Junction—Cove Fort
89 miles/143 km.

Starting out in Castle Valley, part of the racetrack valley eroded in Cretaceous Mancos shale around the San Rafael Swell, this route soon begins to climb through more Cretaceous rocks: thick gray and yellowish gray cliff-forming sandstone that alternates with shaly slopes. Visible from the highway and from the rest stop at milepost 85, these formations belong to the Mesaverde group.

The lower part of the Mesaverde group contains several coal seams. Coal develops where partially decayed plant material accumulates in marshes and swamps, and is later compacted by the weight of sediments deposited above it. Here the alternation of sandstone, shale, and coal results from pulselike retreat of the Cretaceous sea, when periodic advances swept sand across muddy marshes and swamps along the shore. Cretaceous coal in this area is soft or bituminous coal. Probably ignited by lightning, some of it has burned underground, baking overlying rocks to a light brick red color. The Mesaverde group is about 2000 feet thick here, and the Mancos shale about 4000-5000 feet thick. The cliffs are topped with younger rocks that span the Cretaceous-Tertiary time boundary, when dinosaurs and many other forms of life became extinct.

West of the summit are good views of Mt. Musinia or Marys Nipple (10,986 feet), a flat-topped peak capped with Tertiary rocks. The area west of milepost 74 is slivered by many north-south faults; the mountain lies on one of the upthrust slivers. Faults parallel many ridges; a few show up in roadcuts, offsetting rock layers so yellowish-gray Cretaceous sandstone abuts soft, pink Tertiary lake sediments. North-south stretches of the highway run right along some of these faults.

Both Cretaceous and Tertiary rocks are differently expressed on this wetter side of the Wasatch Plateau. Landslides are common. The round-topped scoop below Mt. Musinia is a landslide scar, with rubble from the slide forming typical hummocky landslide topography. Watch for other slides — some of them quite near the road. In places the highway is buckled by slumping — either dropped downward or pushed upward by landslide movement.

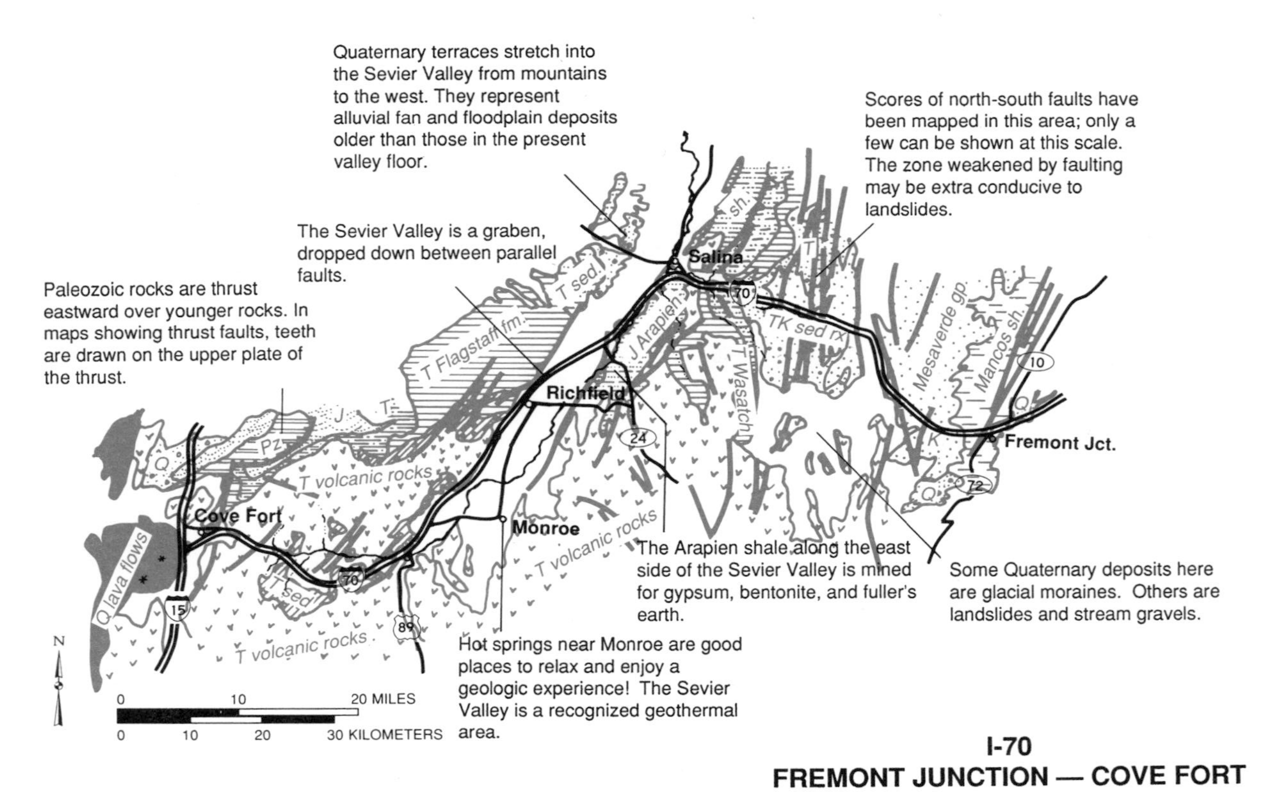

I-70
FREMONT JUNCTION — COVE FORT

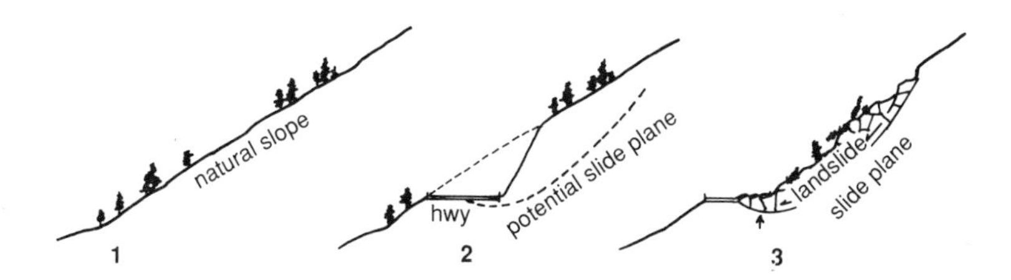

Landslides move along curving slide planes. Those caused by over-steepening of highway cuts may push pavement upward (arrow).

Roadcuts between mileposts 70 and 69 again expose coal-bearing parts of the Mesaverde group. Individual shale and coal layers are discontinuous. Stream channels of cross-bedded sandstone thread through them, as we would expect in a shoreline environment. The discontinuous nature of the coal seams, as well as the many faults, made coal mining difficult here.

More landslides exist farther west; watch for them on both sides of Salina Creek. The many north-south faults in this area add to the likelihood of slides. Springs tend to occur along faults, keeping both rocks and soil wet, heavy, and well lubricated. Earthquakes centered along faults have been known to trigger landslides.

In the big roadcut near milepost 58, tilted Cretaceous sandstone and shale are overlain by younger limey sandstone deposited in a Paleocene lake. A mile or so farther west, very light-colored yellowish-white Arapien shale, a Jurassic formation, appears in the lower canyon walls. Still farther west, Cretaceous and older rocks bend up sharply, reflecting uplift of the Sevier Mountains during Cretaceous time.

A lead-silver mine near milepost 57 is surrounded by these steeply tilted rocks—with, again, Paleocene lakebeds on top. The mines were only marginally successful; ores occurred in narrow sandstone stringers deposited along channels.

At the mouth of Salina Canyon, Tertiary rocks are cut off abruptly by the Sevier fault. The Arapien shale west of the fault is weak and badly distorted by slow flowing movement of its layers of salt and gypsum. The shale also contains selenium. Very few plants tolerate both salt and selenium, so the Arapien shale generally erodes into barren badlands easily recognized by their unusual whitish color, here and there banded with red or yellow. Salt and gypsum are here partly because they are light, and so tend to rise through overlying rocks.

A rockslide in the Green River formation, just south of milepost 59, is easy to see from the highway. Soil development and the growth of vegetation will eventually obscure its features.

Near Salina, Interstate 70 swings southwest along the valley of the Sevier River, a fault-edged depression underlain by Jurassic salt and gypsum, Cretaceous sandstone, and Tertiary lake deposits and volcanic rocks. To the south are small, conical, gray and white peaks of Jurassic Arapien shale, locally mined for salt, gypsum, and very fine clays that develop from volcanic ash. The salt is used for road de-icing, for cattle, and for table salt, the gypsum for plaster, plasterboard, cement, and other products. The clays — bentonite and fullers earth — are used in refining and decolorizing vegetable, animal, and mineral oil. Bentonite is also used to thicken muds employed in drilling for oil and gas.

The Sevier River flows north between the Sevier and Elsinor faults. Its headwaters are in lava-covered plateaus to the south and southeast. Most of its water is now diverted, but in years of plentiful rain it flows out of the mountains and onto the Sevier Desert in the western part of the state. In the years of 1983-1987, it reached Sevier Dry Lake, which became a wet lake for the first time in living memory.

The Pavant Range consists of tilted Cretaceous and Tertiary strata, the latter seen as pink and white bands along the mountain front. These rocks are silty and limy lake deposits closely related to those that decorate Bryce Canyon National Park and Cedar Breaks

National Monument. Paleozoic rocks of the Sevier thrust belt appear on the west side of the mountains, out of view from the Sevier Valley. The southern tip of the Pavant Range is built of Tertiary volcanic rocks, as we shall see.

The Sevier Plateau to the east is covered with volcanic rocks. The Sevier fault, which divides two segments of the Colorado Plateau, runs along the east side of the Sevier Valley, very close to the lower limit of the juniper trees. Displacement along the fault measures thousands of feet. It is still active: recently formed scarps appear at the base of the Sevier Plateau, where they cut across recently formed alluvial fans. Light-colored rocks along the line of the fault are ashflow tuff from Tertiary volcanoes. Many older towns in this area contain buildings made from this volcanic material, easily worked into building stone.

Large gravel quarries mark the valley floor near Elsinor and Joseph. Across the valley similar gravels appear in flat-topped terraces. The water table is close to the surface in the valley, and quarries must constantly be pumped out.

South of Joseph, where the Sevier Valley is filled with volcanic rocks of the Antelope Range, it narrows abruptly. At Sevier the Interstate turns west up the canyon of Clear Creek, with the Tushar Mountains to the south and the Pavant Range to the north. The highest range between the Sierras and the Rockies, the Tushar Mountains consist of a great pile of Tertiary volcanic rocks, the Mt. Dutton volcanics.

These tilted Tertiary rocks were beveled into a mountain pediment, probably in Pleistocene time. With downward erosion since then, the tilted rock of the pediment is well exposed.

Poorly consolidated volcanic ash alternates with layers of breccia; both result from volcanic explosions exceeding any known from historic time.

Steeply dipping layers of dacite breccia and volcanic tuff lie as they fell on the sloping sides of several large volcanoes. The extent and thickness of these rocks show that Tertiary time saw plenty of fireworks here; repeated volcanic explosions exceeded in volume and violence any eruption of historic time. A number of calderas — large circular fault-ringed depressions where volcanoes collapsed into partly emptied magma chambers — are present in these mountains. Although the basins are now full of volcanic rocks, they can be spotted by their circular rings of faults.

Exposed surfaces of some of these rocks are naturally case-hardened by minerals left near the surface as moisture in the rocks evaporated. Interesting and often amusing hoodoos, mushroom rocks, and other figures take shape as weathering penetrates the crusts and wind and rain carve out the softer inside of case-hardened blocks. Columnar jointing, quite prominent in tuff near the mouth of the canyon, formed during cooling and shrinking of the hot volcanic ash.

Some of the massive breccia layers visible near the highway are mudflow deposits formed from newly erupted volcanic ash. Other volcanic sedimentary rocks here are greenish-white lake deposits, in places faulted against pink tuff that weathers into vertical cliffs or conical tepee rocks.

The volcanic rocks remain in view at the highway summit and along Cove Creek Canyon on the west slope. At Sulfurdale, near the mouth of this canyon, small open pit mines once produced sulfur. The pale yellow mineral accumulated near fumaroles, where sulfurous vapors, the ones that smell bad in active hot spring areas, condensed. Sulfur is used in making gunpowder, matches, insecticides, pharmaceuticals, and sulfuric acid, as well as in vulcanizing rubber. Mines here are no longer active, but the hot water and geothermal steam are being used to generate electricity.

Small fault valleys in the Cove Fort and Sulfurdale areas, as well as hot springs, frequent earthquakes, and faults that cut Pliocene and Pleistocene lava flows, indicate that this is a geologically active area.

Cove Fort, built in 1867, is made of local volcanic rocks: basalt for the walls and pink tuff for the fireplaces and chimneys. The basalt came from a lava flow near the junction with Interstate 15.

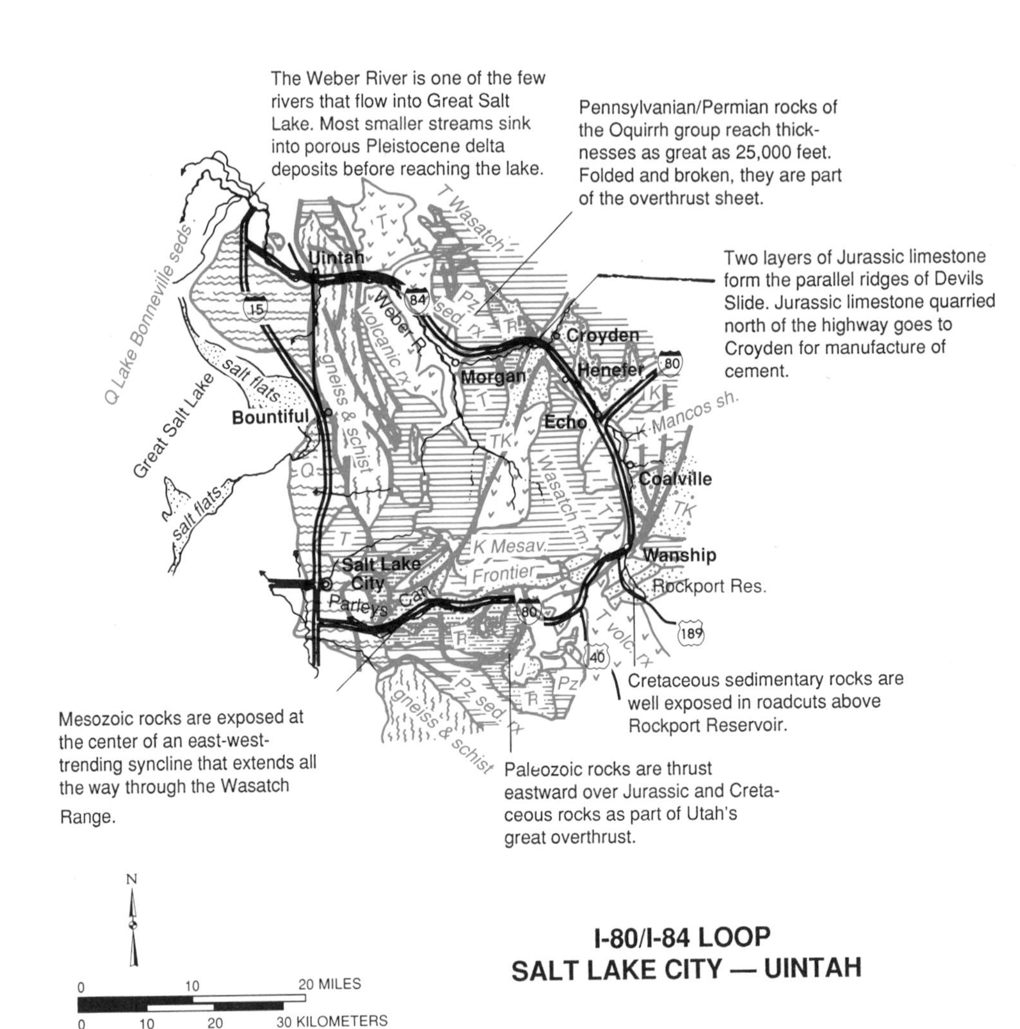

I-80/I-84 LOOP
SALT LAKE CITY — UINTAH

Interstate 80/84 loop
Salt Lake City—Uintah
83 miles/135 km.

Crossing delta deposits that mark the edge of former Lake Bonneville and also conceal the exact position of the Wasatch fault, Interstate 80 plunges into the Wasatch Range via Parleys Canyon. Immense roadcuts display the Jurassic Nugget sandstone, equivalent to the Navajo sandstone of the Plateau country. Quarries here produce attractive pink and tan sandstone for building in the Salt Lake City area. For a time, Jurassic limestone was quarried a little farther east up Parleys Canyon, for Portland cement.

These and other Jurassic rocks extend right through the Wasatch Range in Parleys Canyon. They are folded into an east-west syncline that is cut off westward by the Wasatch fault, the easternmost large Basin and Range fault. Triassic, Permian, and Pennsylvanian rocks, also part of the syncline, appear at higher elevations to north and south, out of sight from the highway. Eastward, Cretaceous units appear near Mountain Dell Reservoir; they are the youngest rocks involved in the syncline.

Near Silver Creek Junction, where US 40 branches off to the southeast, Tertiary volcanic breccia erodes into bizarre shapes — mushrooms and other figures. Note how massive this rock is, with little evidence of regular lava-flow surfaces. Only here and there do we see reddish soil zones baked by hot lava flows. The flows were very thick, formed of particularly stiff magma that hardened and broke up into breccia as it moved.

In the mountains between Silver Creek Junction and Wanship, as in steep mountains everywhere, landslides play a major role in wearing back the slopes. Watch for hummocky surfaces of these slides — all but the newest well vegetated. Large and small, slides can be credited with a considerable amount of highway pavement damage, as well as damage to roadcuts.

Peaks to the south, marked with ski runs, are in the highest part of the Wasatch Range, raised above their neighbors by greater displacement on the Wasatch fault. The peaks are eroded into granitic intrusions that rose during Tertiary time. The intrusions penetrated a west-trending anticline of Precambrian sedimentary rocks that seem to represent a westward continuation of the Uinta Mountains anticline.

Before reaching Wanship we pass through more volcanic breccia from Tertiary volcanoes. Here, too, these dark, massive rocks in some places eroded into mushroom rocks. Elsewhere thin reddish layers were baked by hot lava as it flowed across soil surfaces.

Rockport Reservoir, on the Weber River near Wanship, fills a valley edged with tilted Cretaceous sedimentary rocks, marine sandstone and shale, bent up along the north side of the faulted anticline described above. The rock-fill dam at this reservoir was built in 1955-57. The highway follows the Weber River, which heads in the high eastern peaks of the Uinta Mountains, for the rest of this route. As well as its own water, the river carries water diverted from other Uinta Mountain streams.

Between Rockport Reservoir and Wanship are more landslides, many of them in highway cuts.

A few miles north of Wanship we leave the volcanic breccia and encounter some Tertiary lake deposits, exposed west of the highway. In Tertiary time, during and after uplift of the Colorado and Wyoming Rockies and Utah's Uinta Mountains, large lakes developed in low-lying areas between the ranges. For millions of years, material eroded from the mountains was carried into these lakes. Here we are just at the edge of the lake deposits. East of the highway are Cretaceous near-shore marine sedimentary rocks deposited before the mountains rose, including pale Dakota sandstone, which forms many hogback ridges throughout the Rocky Mountain region.

A tributary stream builds a small delta out into Echo Reservoir. Similar but larger deltas developed along the edge of Lake Bonneville in Pleistocene time.

Echo Canyon is the type locality for the Echo Canyon conglomerate, the locality with which other exposures of the same unit can be compared.

At Coalville, Cretaceous rocks dip about 30 degrees off a northeast-southwest anticline. Hogbacks near the town are in these rocks, which contain the coal mined here. Oil is produced from rocks involved in the Sevier thrust belt just east of Coalville.

Echo Reservoir north of Coalville, also on the Weber River, is filling rapidly with sand and silt eroded from relatively soft Cretaceous and Tertiary rocks.

As the highway approaches the junction with I-84 at Echo, cliffs of thick, coarse Echo Canyon conglomerate come into view ahead. In places more than 3000 feet thick, this rough-looking, bouldery rock, made of alluvial fan gravels washed from the Sevier Mountains, was deposited in late Cretaceous time.

Northwest of Echo, along Interstate 84 through Echo Canyon, high bluffs of this conglomerate edge the highway for several miles. Roadcuts show the conglomerate well, with its boulders up to two feet in diameter. Echo Canyon is also the site of several large landslides, some of them pretty hard on the railroad and highway. The rocks dip toward the canyon, which makes them unstable, and the lower part of the conglomerate rests on clay, which is also unstable. Some of the slides were caused by man's removal of the toes of older slides.

Northwest of Henefer look south, across the highway, for Devils Slide, not a landslide but two limestone ridges tilted into near-vertical

Sedimentary rock units are almost vertical at Devils Slide. This unusual feature is formed by differential erosion of two resistant limestone beds and their softer neighbors.

position. There are something like 4500 feet of Jurassic rocks here: limestone, sandstone, and shale. Limestone quarried north of the highway goes to a cement plant at Croydon.

Between Devils Slide and Morgan we encounter older and older rocks — the same rocks we saw in the syncline in Parleys Canyon, where this roadlog started. Mesozoic rocks give way to shale and sandstone deposited in a gradually deepening Permian sea west of the hingeline.

Mississippian limestone, a single massive, fossil-rich unit of gray marine limestone, appears near milepost 105. West of it are still older Paleozoic strata right down to Cambrian limestones at Morgan. Near milepost 108, horizontally layered Tertiary lake deposits appear south of the highway, but Paleozoic units continue north of the highway. All the Paleozoic strata go up and down like roller coasters between mileposts 109 and 108; farther west they dip less steeply. Everywhere they are bent, folded, and complexly faulted, involved in the mass movement of the great Sevier thrust faults.

Volcanic rocks make up most of the mountains northwest of Morgan. We follow the Weber River through them, riding on flat Pleistocene terraces and lake beds deposited in an arm of Lake Bonneville.

Gravel from Pleistocene terraces is quarried near mileposts 98-97, giving us an inside view of these not-yet-consolidated river and delta sediments.

Dark rocks near and west of the milepost 92 rest stop are the oldest rocks in Utah: Precambrian schist and gneiss about 2.6 billion years old. Here, these rocks are part of the Sevier thrust belt, showing us that the thrust is more than just a near-surface sliding of sedimentary layers. Precambrian sedimentary rocks, also part of the thrust belt, show up a little farther west, on ledges and high slopes and in railroad and highway cuts.

The highway crosses the Wasatch fault near Devils Gate, and there emerges from the mountains. Steep slopes and rushing streams give way to lake deposits, terraces, and old shorelines associated with Lake Bonneville and discussed in Chapter III.

A line of glacier-carved cirques marks the crest of the Wasatch Range. No glaciers exist in Utah today.

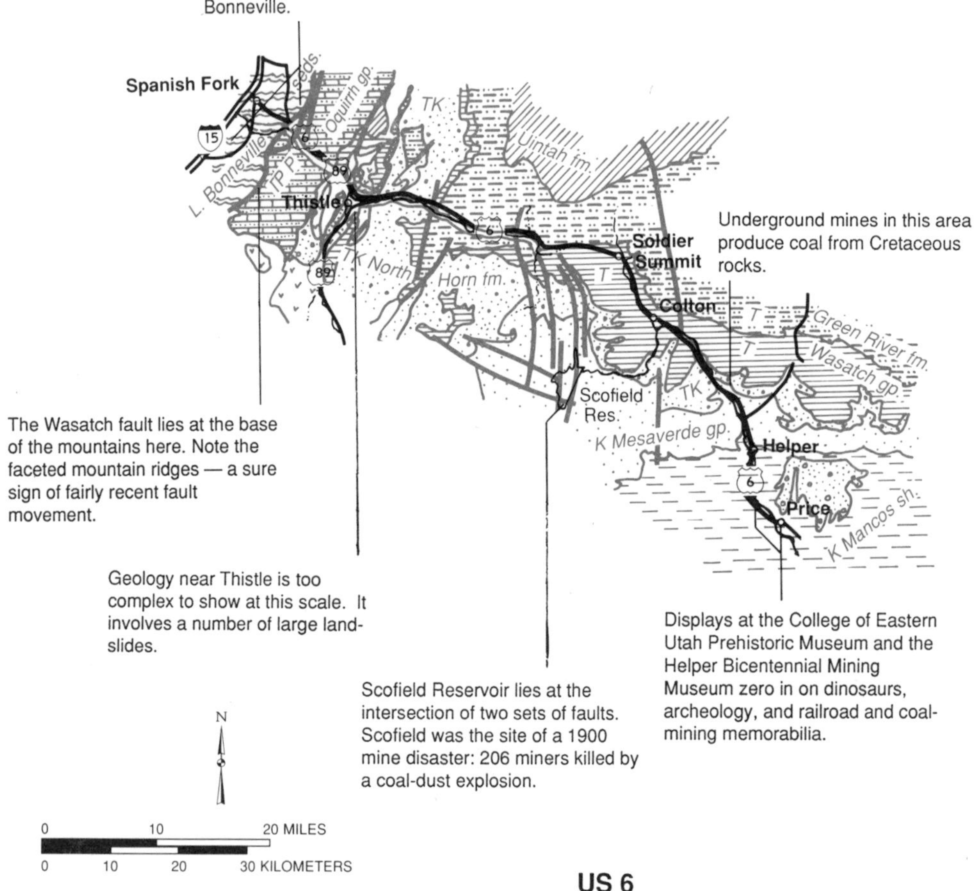

US 6
PRICE — SPANISH FORK

US 6
Price—Spanish Fork
68 miles/109 km.

The College of Eastern Utah Prehistoric Museum in downtown Price displays fossil dinosaurs found in the northern part of the Colorado Plateau. North of Price, US 6 ascends the Price River, heading directly into the Book Cliffs. The lower gray hills and the bases of the cliffs are Cretaceous Mancos shale; cliffs above are a trio of other Cretaceous formations, subdivisions of the Mesaverde group, with sandstones separated by darker shales and seams of coal. The central unit, the Price River formation, is named for exposures in this canyon. Carefully studied and described, these cliffs make up a type section, a sort of yardstick against which other parts of the same formation can be compared.

The gray and yellow-gray Cretaceous rocks were deposited in a coastal environment, with offshore barrier bars, beaches, and lagoons and swamps in which plant matter accumulated, later to become compressed into coal. Some of the sandstones were deposited in stream channels that drained east from the Sevier Mountains of Cretaceous time. In most natural exposures, sandstone layers form cliffs and ledges, shale weathers into slopes. Coal seams are usually hidden by shaly gravel talus.

Coal is mined on a large scale at the junction with US 191, where it is put to use, with no intervening transportation costs, in a power plant right at the mine portal.

The historic mining town of Helper is home for the Helper Bicentennial Mining Museum, which displays relics of the mining and railroad history of the region.

The big roadcuts near the junction with US 191, and others farther up the canyon, expose thin coal beds in the Cretaceous rocks, as well as some of the cross-bedded sandstone deposited by streams from the Sevier Mountains.

On up the canyon there are signs of abandoned coal mines and mill workings, and many other exposures of Cretaceous rocks. The brick red color that shows up here and there, as in the roadcut south of milepost 229, is due to baking of drab Cretaceous strata by underground burning of coal, a natural process probably started by lightning or forest fires.

As the highway continues to climb, we go through younger and younger Cretaceous rocks. Late in Cretaceous time, as the Rocky Mountains began to rise, the Cretaceous sea, the last one to cover this part of the continent, drained away. In broad intermountain basins, lakes developed in early Tertiary time. Fine rock material washed off adjacent mountains into the lakes; today these fine lake sediments make it possible to map the position and extent of these lakes.

In early Tertiary time large lakes occupied parts of Utah, Colorado, and Wyoming. Flagstaff Lake was Paleocene, Lake Uintah was Eocene. —Adapted from RMAG Geologic Atlas of the Rocky Mountain Region.

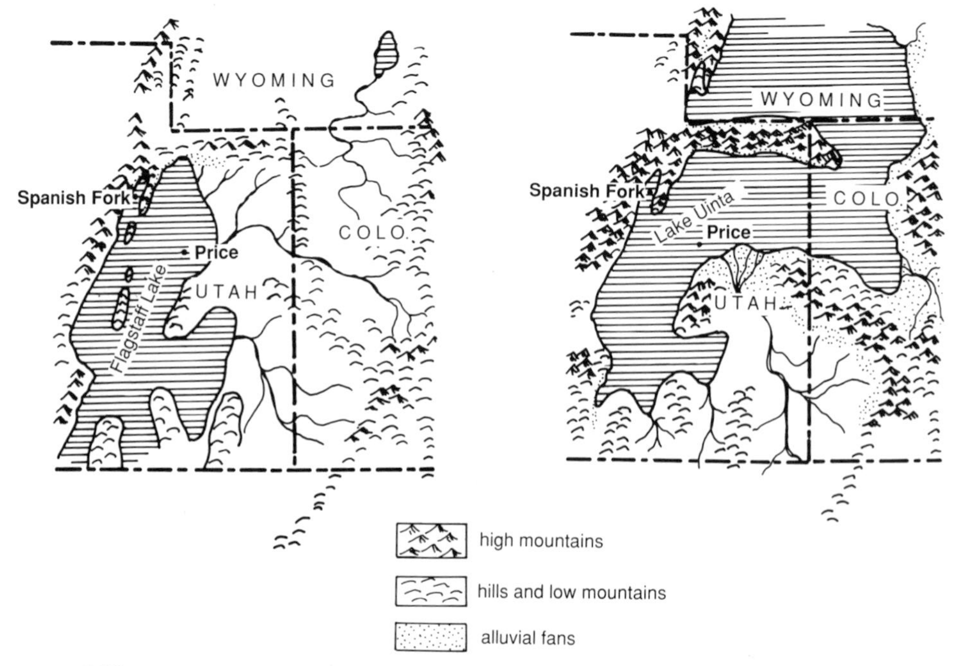

The railroad, zigzagging toward Soldier Summit, crosses and recrosses a hummocky landslide!

We encounter some of the lake deposits near milepost 223. Though all these early Tertiary rocks were once called the Wasatch formation, they have now been divided, as a result of more detailed study, into several separate formations. The oldest and lowest, the Flagstaff formation, forms a sharp-edged ridge on the slope above the Price River bridge at milepost 221. Above it is the Colton formation, reddish mudstone and siltstone named for the nearby town of Colton, representing a delta built out into one of the lakes. The Colton formation is about the same age as Bryce Canyon's sculptured pink rocks, the Claron formation, also deposited in a Paleocene lake.

Near milepost 213 the highway crosses the White River. It gets its whitish color and its name from clays in the Tertiary lake sediments. Note the river's winding meanders — typical of low-gradient rivers — as it crosses the flat valley floor.

Soldier Summit lies close to the intersection of the northern Wasatch Plateau, the Uinta Basin, and the Wasatch Range, geologically a complex region of anticlines, synclines, and faults. However at Soldier Summit itself the geology is simple. At the town and south of it are Paleocene lake deposits that include the Flagstaff limestone and Colton delta deposits; slopes north of the pass are Eocene lake deposits.

Small mine dumps east of the Soldier Summit community mark abandoned ozocerite mines. Ozocerite is a brownish, blackish, or dark green natural wax used in insulation, lubricants, and ink.

Dark, weak, thin-bedded shale and lighter sandstone of the Green River formation represent a change from delta and marsh to an open lake environment.

Strata dip northward here, with predictable results: many landslides in the dipping rock layers along the south side of Soldier Creek. Landslides dam the valley of Soldier Creek at mileposts 208 and 209. Hillsides show scoops where slides have broken away, and trees with twisted trunks indicate that some of the sliding is quite recent.

Unusual fossils have been found near milepost 204: fossil bird tracks in Paleocene lake sediments. West of the rest stop the Green River formation appears as sandstone, gray or greenish gray siltstone, and seams of coal.

The lower part of the Green River formation is well exposed in Baers Bluff near milepost 195. Fossil fish, snails and clams, turtle fragments, and algae-coated logs have been found in limestone and siltstone here. Watch roadcuts for small faults that offset white limestone layers.

Between mileposts 192 and 191, roadcuts show a coarse conglomerate that formed as an alluvial apron along the east side of the Sevier Mountains. Pebbles in the conglomerate include Precambrian quartzite and Paleozoic limestone, thereby showing that the mountains that began to rise early in Cretaceous time were deeply eroded, down to their hard Precambrian cores, by late Cretaceous time. In places the conglomerate weathers into pinnacles or breaks through into high-up windows.

The area near Thistle Junction also sees frequent landslides on both hillsides and roadcuts. Here one of the main culprits is the Cretaceous Cedar Mountain formation, a colorful unit — pastel tones of orange, red, yellow, white, and purple — containing many layers of soft, slippery clay. The highway overlooks destruction caused by the Thistle landslide in the spring of 1983, which dammed Thistle and Soldier creeks. The slide required rerouting and rebuilding of highways US 6 and US 89. The town of Thistle, for a time under 100 feet of water, was not rebuilt. Landslides are one way in which mountains as steep

The lake that backed up behind the Thistle landslide in 1983 destroyed the town and played havoc with the railroad and the junction of US highways 6 and 89. The hummocky lower part of the slide has now been graded, the railroad and highways rerouted.

as these are worn down, one way in which rock material moves from mountain slope to mountain stream, to be carried ultimately to the desert basins west of the mountains.

The contact between Paleozoic and Mesozoic rocks comes between mileposts 180 and 179, where weak red-brown Triassic shale rests on light gray Permian limestone. For the next few miles there are many little fault slices of Permian limestone and Triassic sandstone and siltstone, the Permian units forming gray cliffs and ledges. These faults are at the leading edge of one of the major thrust faults of the Sevier thrust belt. On them Paleozoic and Triassic rocks were shoved eastward across younger strata. Descending to the floor of Thistle Creek, the highway crosses the main fault, with rough gray Permian limestone, part of the 20,000-feet-thick Oquirrh group, steepening slopes to the northwest. Fortunately these rocks are stronger than the Mesozoic rocks near Thistle, and have little tendency to slide. There are, however, some rockfalls here.

Steep alluvial fans at the base of the canyon merge with Pleistocene stream deltas that formerly extended out into Lake Bonneville. The delta sediments now form a wide, flat plain stretching toward Utah Lake. They make up part of the Provo level shoreline (see Great Salt Lake in Chapter III). Higher shorelines of the Bonneville level can be seen on the mountain face. There, too, a line of triangular facets truncates mountain ridges, marking the position of the Wasatch fault, which curves around an embayment in the mountain front.

US 89
Kanab—Panguitch
68 miles/109 km.

Backed by Triassic and Jurassic rocks of the Vermilion Cliffs, Kanab is the gateway to Utah's Grand Staircase — the succession of Vermilion, White, Gray, and Pink Cliffs that edge the southwest corner of the Colorado Plateau. Kanab itself lies on Quaternary stream deposits, pinkish because they are derived from red rocks of the Vermilion Cliffs.

At the base of the cliffs, mostly hidden by fallen rocks and soil, is a narrow outcrop band of Chinle formation, predominantly fine bluish and purplish shale and mudstone well laced with volcanic ash from explosive volcanism in Triassic time. Almost everywhere the unit weathers and erodes easily; here it has eroded off the wide bench that extends southward from Kanab.

Above the Chinle formation, forming the lower part of the Vermilion Cliffs, are Jurassic formations originally deposited on river floodplains and deltas or in coastal dunes. The Wingate sandstone, a Jurassic dune deposit, forms the high, sheer rampart about halfway up the cliffs. Pale rocks at the top of the cliffs, not visible from Kanab itself, are the Navajo sandstone, also Jurassic, one of the great scenery-makers of the Plateau country.

Under the steady wear of erosion the Vermilion Cliffs recede northward at a rate of a few inches per century. Since the strata dip very gently northward here, the cliffs are also becoming gradually lower.

At the mouth of Kanab Creek's canyon, near milepost 67, the river terrace, fine, even-grained, reworked sand from the Navajo sandstone, is deeply gullied, a feature that may be due to introduction of cattle and sheep about 100 years ago. Grazing reduces the amount of vegetation and eventually destroys the spongy, humus-rich soil that normally absorbs rainwater and snowmelt.

From milepost 69, glance upstream at the long, straight canyon of Kanab Creek. The two sides of the canyon don't quite match because it lies right along a fault. North of the bridge is the Navajo sandstone, easily recognized by its buffy white to pink color and its sweeping, large-scale, sand-dune cross-bedding. Because of the uniformity of its

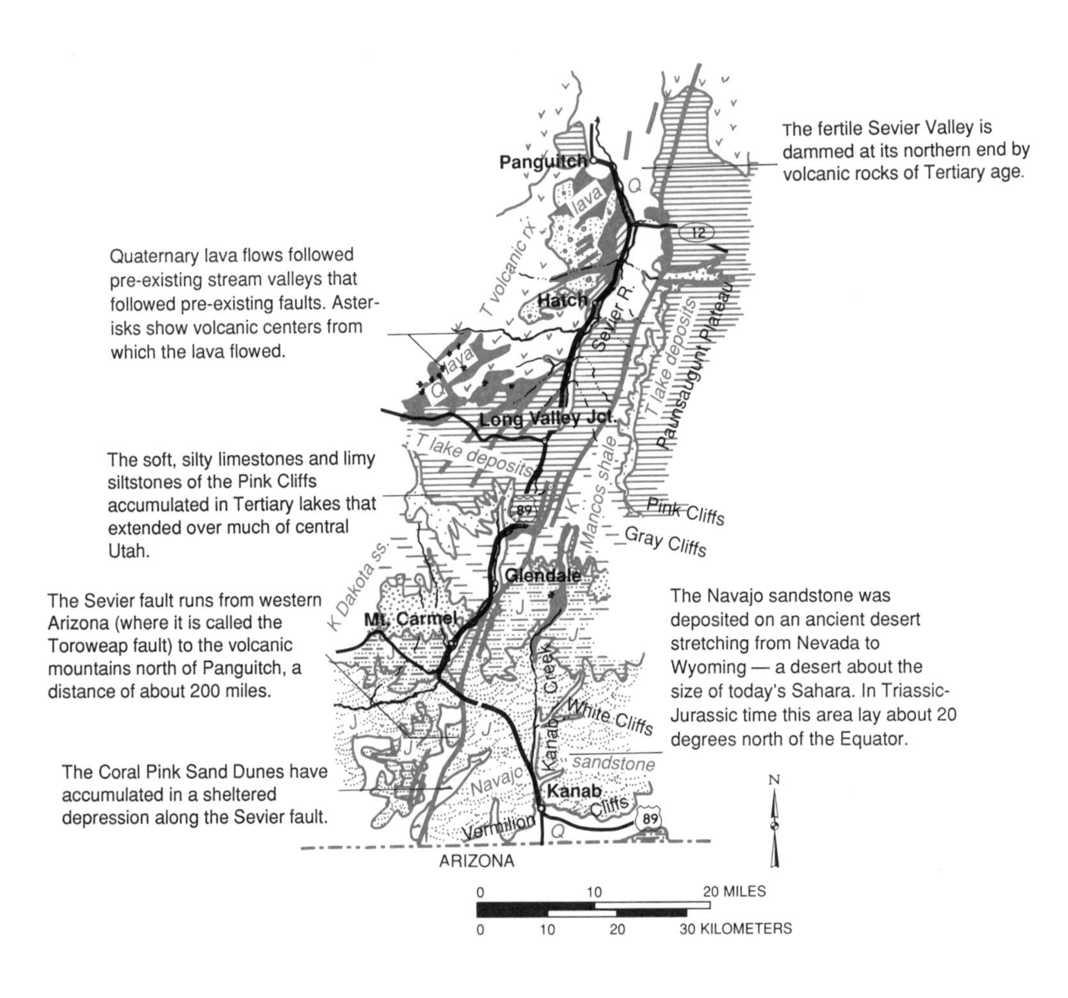

US 89
KANAB — PANGUITCH

Gully erosion in soft stream deposits, seen all over the Southwest, began after 1880. A tributary gully at top center has disgorged a small mudflow. With higher runoff both the mudflow and the collapsed rubble from the gully wall, far right, will be swept away by the main stream.

wind-sorted grains, and because of the calcium carbonate that cements them firmly together, the Navajo sandstone forms another rank of prominent cliffs — the White Cliffs of the Grand Staircase — which will come into view as we climb out of the Kanab Creek drainage and approach Mt. Carmel Junction.

Near the highway we can see many features of this rock, particularly its prominent cross-bedding. In places, especially above the pools near mileposts 71 and 72, the long, diagonal cross-beds are replaced with short looplike swirls formed as sand slumped or avalanched down steep dune faces. There seems to be one particular level, representing one particular time, at which a lot of slumping took place simultaneously. Some geologists suggest that an earthquake may have been responsible. An unusually severe or long-lasting windstorm would probably do as well.

North of milepost 73, the highway travels through an area of coral-colored sand hills, their fine sand derived second-hand from the Navajo sandstone. From the divide at milepost 77, turrets and cliffs of Zion National Park are in view to the west. Zion Canyon's monumental walls are also carved in Navajo sandstone.

Coral Pink Sand Dunes State Park, west of mileposts 77-78, is a more fully developed sand dune field, a good place to check out the behavior of modern slumping sand. Dunes have collected in a topographic depression along the Sevier fault, a major normal fault

On this hillside, light-colored Navajo sandstone alternates with red mudstone, showing that dunes more than once swept across river floodplains, just as modern dunes periodically encroach on the floodplain of the Nile.

separating two sections of the Colorado Plateau: the Markagunt and Paunsaugunt plateaus. Do not confuse this fault with the thrust faults of the Sevier thrust belt farther north.

The highway crosses the Sevier fault near milepost 79. To the east, the Navajo sandstone is high above the highway; to the west the same formation appears below highway level in the canyon of the East Fork of the Virgin River. Displacement along the fault, which extends north

Slumping on steep downwind faces of sand dunes created looplike cross-bedding in the Navajo sandstone. Straight, even cross-bedding below indicates the direction of prevailing winds that piled up these dunes — from left to right in this photograph.

and south for more than 200 miles, is several thousand feet.

Mt. Carmel Junction and Mt. Carmel lie close to the Sevier fault, with dramatic cliffs of Navajo sandstone to the south. To the northwest, by contrast, are low hills of soft, limy, gypsum-bearing Jurassic siltstone, quite easily eroded. A large mass of white gypsum appears near the highway just west of Muddy Creek and Mt. Carmel. Seams of low-grade coal mark some of the hills north of Mt. Carmel.

The highway proceeds up Long Valley, which closely follows the Sevier fault. All these rocks evidently made the town of Orderville rock-conscious; it boasts several gem and mineral stores.

Near Orderville, the Navajo sandstone and the Carmel formation above it dip north more and more steeply. Eventually they plunge underground, leaving younger Jurassic siltstone and sandstone and some gray Cretaceous rocks at the surface. Because of displacement along the Sevier fault, we first encounter Cretaceous strata on the west side of the valley, just north of Orderville. East of the valley they don't show up until Glendale. Where they have eroded into cliffs, they form the Gray Cliffs of Utah's Grand Staircase.

Now surrounded by farms and orchards, Glendale was originally a coal-mining town producing coal from seams in Cretaceous rocks. North of town, Jurassic gypsum gives the hills a strange gray cast.

The rest stop near milepost 95 is in a landslide region where springs lubricate clay-bearing rocks. Landslides can often be identified by their hummocky topography, as is the case here. A mile north of the rest stop, look west across the valley to Little Bryce, where erosion of Tertiary sediments has created one of the breaks typical of the edges of the Markagunt Plateau.

The highway climbs through a lava dam and into the upper part of the East Fork's valley between mileposts 96 and 97. The natural dam and the sediments it held back are responsible for the fertile fields farther north.

The Sevier fault is still with us, at the base of the Paunsaugunt Plateau to the east. Bryce Canyon lies on the other, eastern side of this plateau. Rocks that surface the Paunsaugunt Plateau appear at highway level west of the fault. The soft pink lake deposits are less spectacular here where the landscape is flatter than it is near Bryce Canyon.

Utah 14 to Cedar Breaks branches off at Long Valley Junction at milepost 104. Though not as large as Bryce, Cedar Breaks is another spectacular example of headward erosion into Tertiary lake deposits, erosion that in bits and pieces forms the Pink Cliffs.

Erosion of Tertiary lake deposits of the Claron formation previews the glories of Bryce Canyon National Park and Cedar Breaks National Monument.

North of Long Valley Junction our highway gradually descends the valley of the Sevier River, still following the Sevier fault. Here the river's gradient is low, and it meanders across a wide floodplain. Sunset Cliffs, with more Bryce-like erosion of the Claron formation along the west side of the Paunsaugunt Plateau, can be seen from milepost 110.

A dark basalt lava flow near Hatch, visible ahead from milepost 118, flowed east down a valley, forcing the stream aside. The stream then eroded new cuts that removed ridges on either side, in a sort of reversal of topography we see frequently in lava-flow areas: Valleys become long, narrow lava ridges, and original ridges become new valleys.

East of milepost 121 the Sevier fault shows up well: Pink lake sediments of the Claron formation east of the fault abut much younger lava flows west of the fault. The fault crosses Utah 12, the road to Bryce Canyon, 2.5 miles east of the intersection with US 89.

Notice the river terraces about 25 and 50 feet above the river north of this intersection. Terrace gravels, showing up in roadcuts, include rocks washed from volcanic highlands to the west. Cobbles and

The Sevier fault crosses the center of this photograph and abruptly cuts off pink lake sediments of the Claron formation. Darker rocks west of the fault are lava flows.

pebbles are rounded and sorted, sure signs of transportation and deposition by water.

Views from milepost 129 reveal the Sevier River's floodplain, with marks of its former channels. In places the river has been artifically channeled to protect surrounding fields from changes in its course.

As the highway enters Panguitch, you see good views west to the volcanic uplands of the Markagunt Plateau.

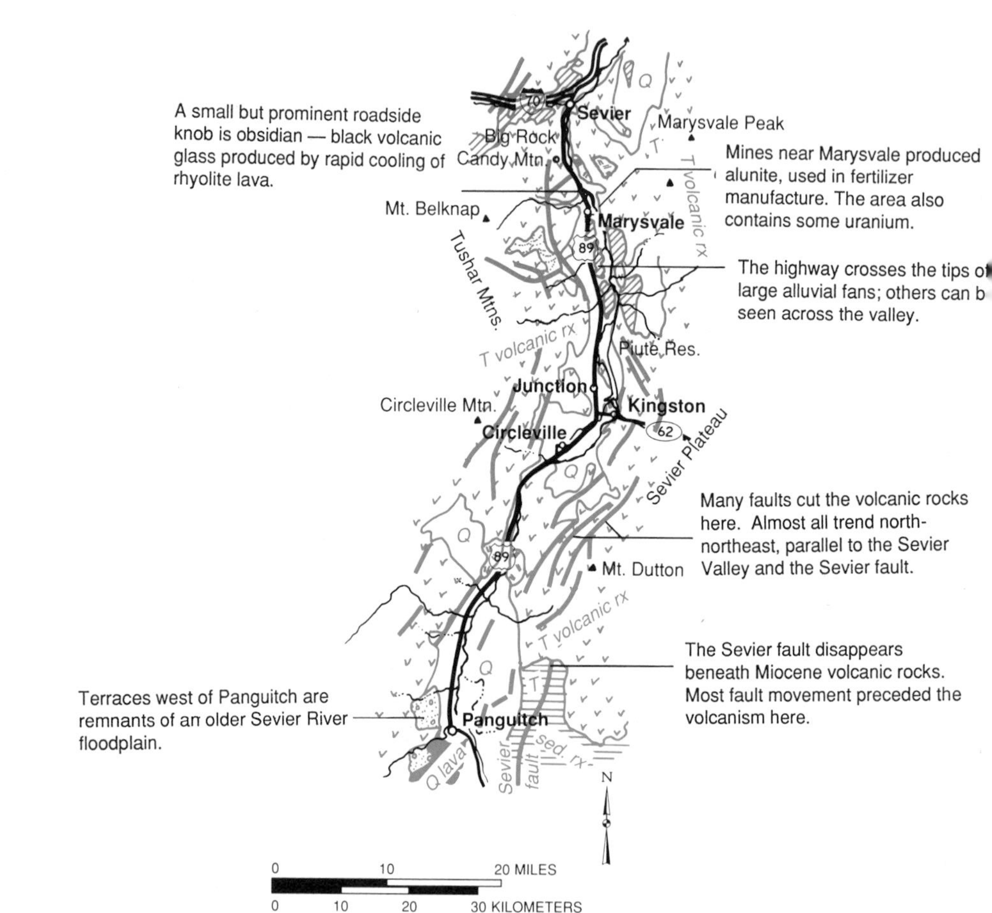

US 89
PANGUITCH — SEVIER

US 89
Panguitch—Sevier
60 miles/97 km.

West of Panguitch are some of the dark volcanic uplands of the Markagunt Plateau — a scene of explosive volcanism in Tertiary time. To the northeast, Tertiary volcanic rocks cap the Mt. Dutton highland, which is separated from the Paunsaugunt Plateau by a zone of intense faulting. Both the Mt. Dutton region and the Paunsaugunt Plateau were lifted by movement along the Sevier fault, which runs along the base of the juniper-pinyon forest on the east edge of Panguitch Valley.

Northeast of Panguitch are good views of jagged, tessellated cliffs eroded in great piles of volcanic rock. River terraces and alluvial fans scallop the east side of the valley. The highway rides on similar alluvial fans on this side of the river. Because this part of the Sevier River's valley is plugged at its north end by Tertiary volcanic rocks, the river here has a very gentle gradient, and wanders back and forth across a wide, flat floodplain underlain by lake deposits. Oxbow lakes and sinuous ponds mark abandoned river channels. In places, particularly at its north end, the valley floor is marshy.

North along the line of the highway at milepost 139, in the Tushar Range, is Circleville Mountain, a partly eroded Miocene volcano. Mt. Dutton, the remnant of another Tertiary volcano, rises to the northeast. Both of these mountains were composite volcanoes or stratovolcanoes akin to Japan's Fujiyama and America's Mt. St. Helens, composed of alternating sequences of volcanic ash and lava flows, with lots of volcanic breccia mixed in. Made up of broken pieces of volcanic rock, breccia forms when movement of thick, sticky lava breaks up its own cooling crust, or when older volcanic rock is broken up during violent volcanic explosions.

Eroded pediments, as well as alluvial fans, edge the valley here. Though they often look like alluvial fans, pediments are products of erosion, carved into the solid rock of the mountains. Like many

The granitic magma of the Spry intrusion cooled slowly deep below the surface, allowing visible crystals to form.

pediments, these are covered with a thin veneer of stream-deposited gravel. Pediments in this area are as much as 100 feet above the present stream level. Most of them have been cut off at their toes as the present stream swings back and forth on its valley floor.

North of milepost 146, our route encounters some massive granite, part of the Spry intrusion, of Oligocene age. The magma from which it formed built up in great underground domes. In places heat from the intrusion altered overlying limestone, changing it to marble.

In contrast to the wide valley in which we've been driving, the Sevier River north of milepost 152 is constrained into a narrow canyon walled with volcanic rocks. Road and river crowd together here. The volcanic centers, once active craters, are not far from this highway: the great volcanoes of Mt. Dutton, now to the east, and the Tushar Range to the north.

Most of the light-colored rock exposed in the canyon is ashflow tuff, formed from incandescent volcanic ash that shot down the flanks of a volcano and welded itself together when it touched down. Such tuff erodes into steep and often pinnacled slopes like those seen in this canyon — in places light buff in color, elsewhere weathered to a dark brown. Tuff containing lots of fragments of broken lava may have originated as mudflows.

Broken by faults and joints, both tuff and breccia break up into rockslides that spill down the canyon walls. A lava flow exposed in a roadcut near milepost 155 lies on a thick red layer of soil baked by the hot lava. Volcanic rocks completely cover the Sevier fault here.

North of Circleville the valley widens out again and the Sevier River meanders nicely across a floodplain that was once a lakebed. Mt. Belknap, Delano Peak, and City Creek Peak crown the Tushar Range, a huge pile of Tertiary volcanic rocks that makes up the tallest mountains between the Rockies and the Sierras.

On the northern part of the Sevier Plateau are more Tertiary volcanoes, including a caldera formed by explosive eruption and collapse of the volcano into its partly emptied magma chamber. Here the Sevier fault, which resurfaced near Junction, again disappears beneath volcanic rocks.

Close to the heart of the now deeply eroded Mt. Belknap volcano is the picturesque town of Marysvale. Lead-zinc mining in the Tushar Range put Marysvale on the map. During World Wars I and II, decomposed volcanic rock was mined as a source of alunite, used to make alum and fertilizer. A town called Alunite, south of Marysvale, lay near another source for this material. More recently, the Antelope Hills north of Marysvale have produced uranium; ores occur where a Tertiary intrusion broke up surrounding volcanic rocks, allowing penetration of hot, mineral-bearing solutions.

About 1 1/2 miles north of Marysvale, a cliff of obsidian borders the highway on the west. Obsidian is volcanic glass that forms from lava that hardens as it cools, but does not crystallize. The obsidian cliff is the edge of another caldera.

Dips shown by these volcanic rocks are original or primary dips, formed as lava flows, layers of breccia, and thick sheets of volcanic ash built up the steep sides of stratovolcanoes.

Soft, brightly colored slopes of Big Rock Candy Mountain are composed of decomposed volcanic rock altered by hydrothermal (hot water) solutions. The yellowish color is limonite, an oxide of iron.

Tall breccia pinnacles near the milepost 185 rest area mark more volcanic rocks of the Mt. Belknap volcano. North of these pinnacles similar rocks have been altered to yellowish clay as hot water and steam, well supplied with volcanic acids, rose through them — the same type of hydrothermal alteration that is responsible for the yellow rocks of Yellowstone National Park.

Big Rock Candy Mountain, complete with a lemonade spring, is another area of hydrothermal alteration. Oxidized iron, limonite, gives the clay and the spring water their rusty yellow color. The rocks are weakened by the alteration process, and erode readily, as you can see. A small gray igneous intrusion in the altered rock across the canyon may have heated and acidified the hot water that caused the alteration.

About five miles north of Big Rock Candy Mountain, both river and highway veer eastward around a mass of rock that looks like breccia. It has been identified as a mudflow deposit from a Mt. Belknap eruption. There is more hydrothermal alteration of the volcanic rocks here, some of it near the junction with I-70 near Sevier. Just north of this junction, visible from either highway, are high pediments carved in tilted greenish and pinkish layers of volcanic ash.

US 89 between Sevier and Salina is described, from north to south, under I-70 Fremont Junction to Cove Fort.

This pinnacled rock mass, composed of volcanic tuff and fragments of lava, is part of a mudflow formed in Tertiary time as a result of a volcanic eruption.

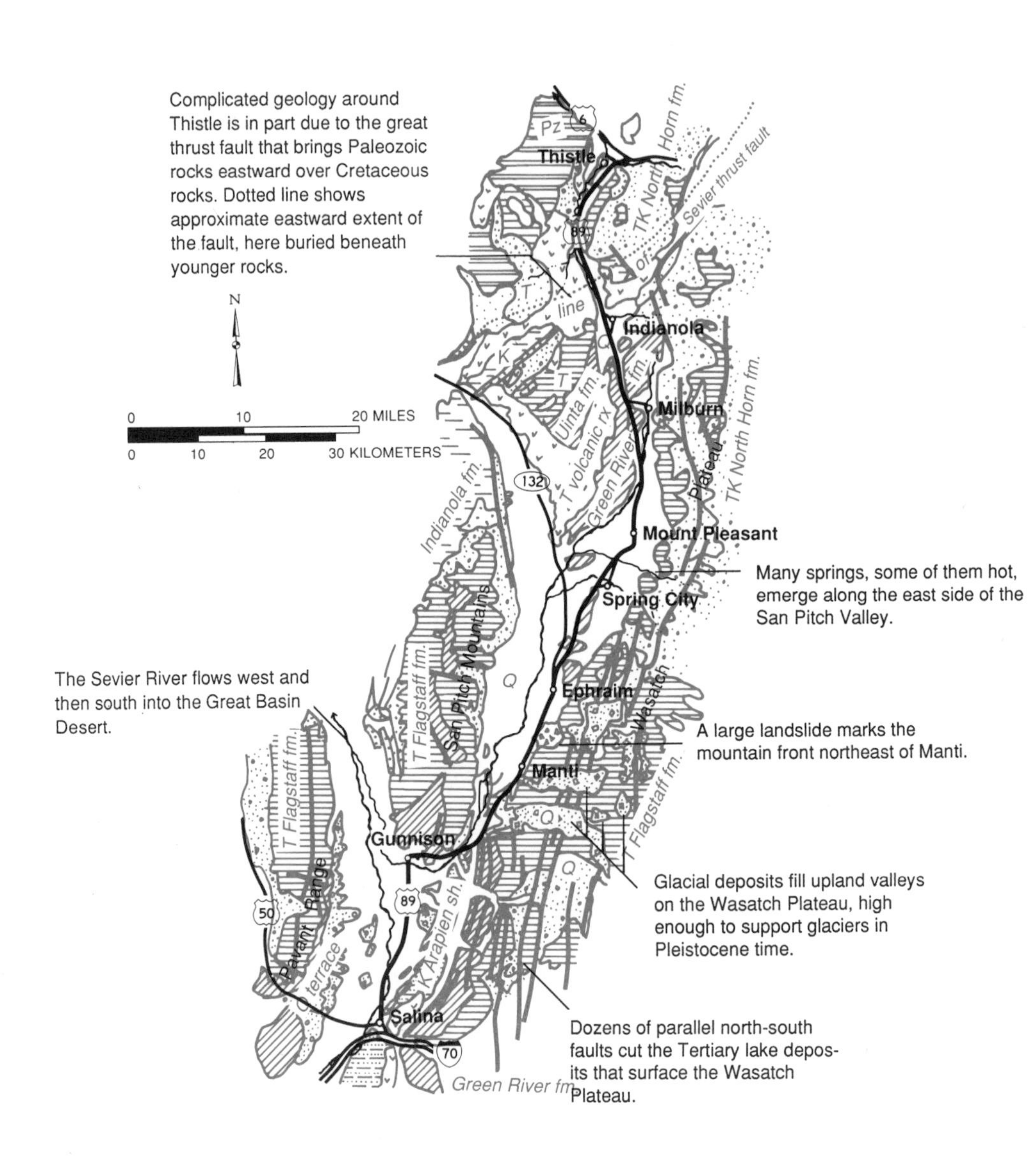

US 89
SALINA — THISTLE

US 89
Salina—Thistle
88 miles/140 km.

Heading north from Salina, this highway gives good views of the ranges that border the Sevier Valley. Those to the west, the Pavant Range, bear the stripes of sedimentary rocks, mostly Cretaceous and early Tertiary strata. Pink and white bands mark lake deposits similar to those at Bryce Canyon and Cedar Breaks farther south, and were deposited in a large lake that spread between the newly risen Wasatch Range and Rocky Mountain ranges in Wyoming and Colorado.

The west side of the Pavant Range, out of sight from here, exposes Cambrian quartzite and limestone pushed eastward over Jurassic sandstone by movement on the Sevier thrust faults, formed in Cretaceous time. Faults of the Sevier thrust belt should not be confused with the Sevier normal fault, which edges the Paunsaugunt and Sevier plateaus farther south. The Pavant Range lies right on the forward edge of the Sevier thrust fault. It also lies right on the Paleozoic hingeline, so its Cambrian rocks thicken rapidly westward.

The Wasatch Plateau to the east is composed of similar Cretaceous and Tertiary rocks, all much faulted, with arrays of parallel north-south faults slicing the whole plateau into long, narrow slivers. Mt. Musinia, a white-topped peak rising well above the rest of the Wasatch Plateau, is an upfaulted remnant of lake-deposited limestone again about equivalent in age to that at Bryce Canyon and Cedar Breaks.

Cuestas along the plateau front are eroded in Eocene rock, the Green River formation, sliced by faults and tilted like so many slices of a loaf of bread. Jurassic Arapien shale, a weak white and light gray rock mined west and south of here for salt and gypsum, shows between some of the cuestas.

North and south of Salina, the valley of the Sevier River is a geothermal area. A warm spring, temperature less than 122° F (50°C), is just north of Salina. Other warm springs occur farther south near Monroe and Sevier, and to the north near Sterling and Manti.

In Pleistocene time an arm of Lake Bonneville reached into this part of the Sevier Valley and almost as far south as Sigurd. The fertile valley floor, so flat that the Sevier River seems at a loss to know which

Cuestas east of Gunnison expose tilted shale and limestone of the Green River formation, deposited in an early Tertiary lake.

way to go, is a legacy of that lake. Lake Bonneville shorelines are visible on mountain slopes near Gunnison.

At Gunnison we leave the Sevier River, which curves gradually westward here. After following an east-west fault that crosses the valley right at Gunnison, our highway heads up the valley of the San Pitch River. To the southeast, early Tertiary (Paleocene) lakebeds can be seen behind hogback slices, essentially landslide blocks, of the slightly younger Green River formation. The Paleocene rocks also appear in roadcuts, badly fractured, along with grayish and yellowish Arapien shale, a Jurassic unit. Gray, yellow, and pink badlands of Arapien shale appear in Arapien Valley to the southeast.

Cuestas of Green River formation — landslide blocks similar to those farther south — help contain the water in Gunnison Reservoir. Outcrops and roadcuts near the reservoir expose this rock, which varies from sand and silt to lake limestone. Near and north of the reservoir, the cuestas are half buried beneath the flat floor of the San Pitch Valley. One of them forms the hill on which Manti Temple stands. Another, about two miles north of town, provided limestone for the temple from the Green River formation. The limestone is oolitic — made of small spheres that look like fish eggs but that are really formed as sandlike grains roll around in agitated water. Some older homes in Manti are made of the same material. The formation contains fossils of fish, alligators, turtles, and other inhabitants of the early Tertiary lakes.

Mountains on either side of the valley are still young and steep. As a result, landslides, rockfalls, and earthflows are common. The San Pitch Mountains, now to our west, are made up of tilted Cretaceous and Tertiary rocks. The Wasatch Plateau to the east is similarly composed, and in addition is faulted along this side. Between Manti and Ephraim, landslides along the edge of the Wasatch Plateau have stripped surfaces of the Eocene lake deposits on which they slid. Partly because of the many faults, there are numerous springs along the east edge of the valley, a boon for early settlers.

The valley narrows north of Mt. Pleasant, but is still bordered by ridges of Paleocene and Eocene sedimentary rocks. North of Fairview, the highway climbs over these rocks and into the drainage of Thistle Creek. Volcanic rocks, partly reddish breccia, border the highway

A few partly demolished buildings still stand in and near Thistle, well below the high-water line that marks surrounding mountainsides.

During the spring and summer of 1983, the slide-dammed lake at Thistle cut this beach cusp (arrow) about 100 feet above the town. Similar but more pronounced shorelines were cut by Lake Bonneville in Pleistocene time.

between Indianola and Birdseye, while to the north and west, higher mountains appear: dark summits of late Paleozoic (Pennsylvanian and Permian) limestone. These rocks are part of the belt of Paleozoic rocks that were pushed eastward in Cretaceous time as the Sevier thrust belt. High on the mountains, close to their crest, are cup-shaped cirques left by little Pleistocene glaciers.

Between Milburn and Indianola we cross a drainage divide, leaving the headwaters of the San Pitch River and entering the drainage of Thistle Creek. This creek is a tributary of Soldier Creek, whose waters flow into Utah Lake near Spanish Fork. North of milepost 272, Thistle Creek has cut away its banks and then taken a shortcut, leaving its former course as an oxbow lake.

Our highway follows Thistle Creek as it runs diagonally across the general northeast trend of the rock strata. North of Indianola both Cretaceous and Tertiary sedimentary rocks are covered with reddish volcanic breccia.

A short distance downstream from Birdseye, dead trees along Thistle Creek mark the extent of a lake that formed here in the spring of 1983, when a large landslide dammed the canyon of Soldier Creek below the town of Thistle. The slide occurred after heavy rains, when rainwater not only lubricated soft shales near the town, but also added to the weight of rocks and soil above them. The slide did not hit the town, but water rising behind the dam soon inundated the little community. Because of the danger that the lake would overtop the slide dam and then rapidly cut down through the loose boulders and gravel and mud in the slide, releasing a dangerous and destructive downstream flood, the Army Corps of Engineers tunneled through the slide to drain the lake gradually.

The junction of US 89 with US 6 was moved from the canyon floor to a point much higher on the mountainside. A viewpoint north of the new roadcuts on US 6 provides an excellent view of the slide. Slide-prone rocks caused numerous smaller slides in these roadcuts even as the new highway was being built.

Dinosaur bones have been found in Cretaceous rocks of the Cedar Mountain formation near Thistle.

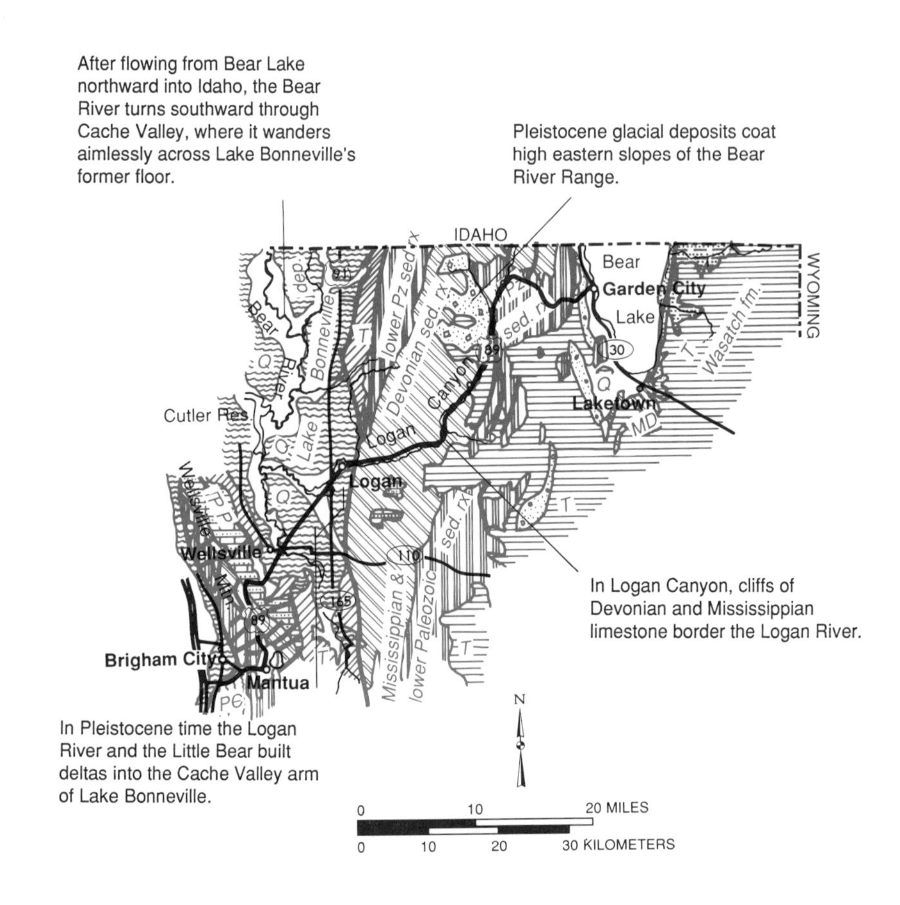

US 89
BRIGHAM CITY — GARDEN CITY

US 89
Brigham City—Garden City
64 miles/103 km.

Wellsville Mountain east of Brigham City is a long, upward-faulted ridge of Precambrian metamorphic rocks covered with east-dipping Paleozoic sedimentary rocks. An essentially complete sequence from Cambrian to Permian appears as US 89 cuts eastward and then northward through the range.

As with other parts of the Wasatch Range, the western face of Wellsville Mountain is defined by the Wasatch fault, with the mountain on the east side of the fault lifted thousands of feet relative to the Great Basin region west of it.

Precambrian metamorphic rocks surface the curving mountain front to the south. They consist mostly of phyllite, a shiny, silvery type of schist that glistens with flat, shiny grains of mica and chlorite.

Above the Precambrian rocks and forming most of the mountain escarpment near Brigham City is the Brigham quartzite, a resistant, cliff-forming unit originally deposited as sand along the edge of the Cambrian sea, at that time advancing across this area from the west. The Brigham quartzite correlates with sandstones that, known by other names, stretch halfway across the continent, reflecting the gradual invasion of the Cambrian sea. The sandstones become younger eastward, showing that the sea, with its beaches and bars, crept slowly in that direction.

Upper parts of the mountain front are Cambrian too — gray shale and limestone layers deposited as the sea deepened and the source of sand became farther away. Cambrian sedimentary rocks once covered all of Utah; they are still more widely distributed than sedimentary rocks of any other period. Everywhere the sequence is the same: sandstone at the base, then shale, then thick layers of fossil-bearing limestone or dolomite. Dolomite is similar to limestone but with added magnesium carbonate.

The highway turns north near Mantua and soon crosses the contact between Cambrian and Ordovician rocks. It then makes its way obliquely across younger Paleozoic rock layers: Silurian dolomite and Devonian sandstone, conglomerate, and limestone. Like the Cambrian rocks, these units are marine or near-shore deposits. They all thicken westward, reflecting greater subsidence in that direction. Some of them contain well preserved marine fossils: corals, trilobites, bryozoans, sponges, echinoderms, brachiopods, graptolites. The sandstone, limestone, and dolomite are fairly resistant and form cliffs and ledges on the mountainside, whereas shaly units form slopes and benches. The highest cliffs are thick Mississippian limestone, deposited in fairly deep water far from land. Like the Cambrian sandstone, this unit reaches halfway across the continent, and correlates with thick limestone formations of Montana, Wyoming, Colorado, and Arizona. Unlike the Cambrian sandstone, the limestone was deposited in an open sea with little or no land-derived sand or silt.

The east slope of Wellsville Mountain, gentler than the faulted western slope, is in Pennsylvanian and Permian rocks — limestone and sandstone that reflect changing topography in various parts of Utah. Submarine ridges restricted free circulation of seawater, and basins collected unusual types of sediments that include salt and phosphate.

All of the Paleozoic rocks we have been seeing are part of the Sevier thrust belt. They moved about 40 miles east during the building of the Sevier Mountains in Cretaceous time.

East of the summit of Wellsville Mountain, the highway descends into Cache Valley, a true graben, dropped between the East Cache fault and the Bear River Range. Cache Valley was covered in Pleistocene time by an arm of Lake Bonneville, Great Salt Lake's ice age ancestor. Old Lake Bonneville shorelines mark the surrounding mountains. For a time, the lake's outlet was northward through Cache Valley into Idaho. Now, the Bear River enters the valley from the north, wanders in irregular meanders across the flat lakebeds of the valley floor, then escapes westward via the narrow pass between Wellsville Mountain and Gunsight Peak farther north. Ultimately the Bear River reaches Bear River Bay in Great Salt Lake. We'll learn more about this river's erratic course at Bear Lake.

At Logan, the highway climbs onto a gravel delta formed in Pleistocene time by the Logan River, which at that time emptied into the Cache Valley arm of Lake Bonneville. Gravels of the delta can be seen near the mouth of Logan Canyon. The delta lies at the Provo level of Lake Bonneville; wavecut shorelines of the Bonneville level mark the mountain front above the delta.

Delta deposits hide the exact location of the East Cache fault, but the steep mountain face shows its approximate position. Displacement on this fault may be as great as 12,000 feet. Where it appears at the surface, it is nearly vertical — a typical normal fault. However,

Ordovician limestone is an almost constant highway companion between Logan and Ricks Spring.

seismic studies, in which small earthquake waves from manmade shocks are reflected off underground rock layers, show that the fault curves and flattens out below the surface.

As the highway crosses the Logan River and heads into Logan Canyon, it enters Paleozoic rocks once more—the same sequence that we saw on Wellsville Mountain. Cambrian rocks at the mouth of the canyon are overlain by Ordovician limestone near the Logan City Power Plant, and then by Ordovician quartzite and black dolomite a short distance beyond.

Silurian dolomite, lighter in color, appears between mileposts 375 and 376, with smooth-sloped Devonian rocks beyond milepost 376. In contrast to corresponding rock units of Wellsville Mountain, here the Paleozoic rocks flatten out and even dip westward, between mileposts 381 and 382, in a shallow syncline. Low-angle thrust faults repeat the rocks as well. As a result, the highway remains among these Paleozoic rocks for some distance, going through the same units again and again. Above the highway, cliffs of Mississippian limestone form the upper canyon walls.

A roadside sign near Spring Hollow, between mileposts 378 and 379, points out the gravelly beach terrace that documents the farthest extent of Lake Bonneville up Logan Canyon. A little beyond this point, massive, much-jointed Paleozoic dolomite has eroded into picturesque pinnacles. The Brachiopod Summer Home Sites are named for fossils found in Paleozoic rocks of this area! Brachiopods are marine shellfish having two bilaterally symmetrical shells. They are rare today, but were plentiful in Paleozoic time. Notice the steep profile of the narrow canyon, and the greater amount of vegetation on its cooler, wetter, north-facing slopes.

The rocks near Ricks Spring, which emerges from a cave in Ordovician limestone, are highly faulted and riddled with solution caverns etched by groundwater as it percolated through the limestone. Such water, made acid by absorbing carbon dioxide from the air and soil, slowly dissolves the limestone, carving out underground channels and caverns. In places the strata have been tilted by drag along some of the many faults in this area.

Above Ricks Spring the Logan River flows in a canyon cut deeply into Ordovician quartzite. Before excavating this cliff-walled trench, the river for some time meandered across the surface of the hard quartzite; its meanders are preserved as tight loops in the present canyon.

Rounded slopes farther along the highway suggest softer, more easily eroded rocks. Their limy shales and shaly limestones belong to the Wasatch group, much younger and weaker than the rocks we've

A strangely marked rock surface tells of burrowing worms, mollusks, and other marine animals that tunneled in the Ordovician sea floor in their search for food. Lens cap, two inches in diameter, gives scale.

Ricks Spring flows from a cave in Ordovician limestone. Its maximum flow, over 30,000 gallons per minute, comes in late spring and early summer, its source the melting snow of surrounding mountains. The spring dries up completely in winter, when there is no snowmelt.

been seeing. More widespread east and southeast of here, they were deposited in Paleocene lakes between the newly risen Wasatch Range of Utah and the Wyoming and Colorado Rocky Mountains. Between mileposts 402 and 403, roadcut exposures reveal many small faults that break and offset the limestone and shale layers, further adding to the ease with which they erode.

The highest part of the Bear River Range was glaciated in Pleistocene time. The highway crosses bouldery glacial deposits near the summit of the range.

Bear Lake comes into view west of the summit, its brilliant turquoise waters contrasting with surrounding sage-covered slopes. Several highway overlooks make good stopping points for viewing surrounding geology. Bear Lake lies in a graben between two faults. Rocks west of it, here in the Bear River Range, are Paleozoic, as we've seen. Those east of the lake are almost entirely Mesozoic, with upward-faulted Nugget sandstone (Jurassic) forming the resistant ridge that confines the east side of the lake. Reddish Tertiary lake deposits — the Wasatch group again — lap up onto these Mesozoic rocks, capping the distant hills.

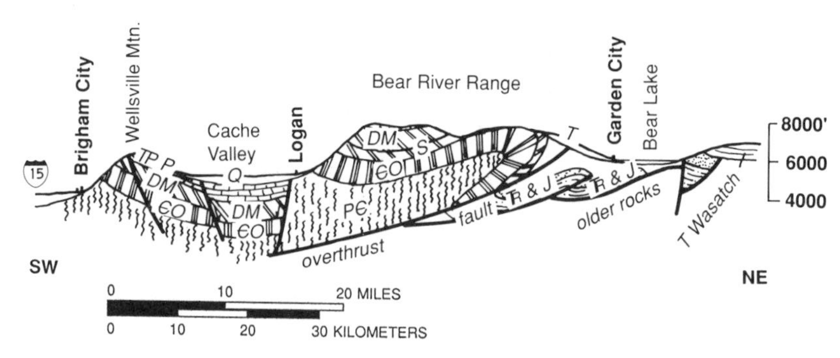

Section parallel to US 89 Brigham City to Bear Lake

Though popular thought would have Bear Lake a remnant of Lake Bonneville, it is too high for that. It was, however, larger in Pleistocene time than it is now. It drains north into Idaho via the Bear River, which then, as we've already seen, does an about-face, turning south into Cache Valley and then west and south into Great Salt Lake. The brilliant turquoise blue of Bear Lake results from its shallowness and its white sand bottom. The lake is edged with large alluvial fans; small deltas mark the mouths of tributary canyons.

The view southward includes the distant Uinta Mountains, a high, rugged range that is basically a single large faulted anticline. (This range is discussed under Utah 150 Kamas — Wyoming.)

Highway 89 descends to the edge of Bear Lake at Garden City, and then follows the western lakeshore north into Idaho.

Cone-shaped rockslides in Paleozoic strata mark the mountain front north of Provo. Note the funnel-shaped area feeding the tops of the slides.

US 189
Provo—Wanship
60 miles/97 km.

Provo's newest subdivisions are moving up — up onto the old Lake Bonneville delta of the Provo River: nice, nearly level homesites overlooking the rest of the town. The delta is part of the Provo shoreline; above it are traces of the older, higher Bonneville shoreline, as well as a few intermediate shorelines. (For more on these shorelines, see Great Salt Lake in Chapter III.) The delta, and with it most of the subdivisions, straddles (and hides) the Wasatch fault. Truncated mountain spurs show that the fault has moved fairly recently, before historic time but not all that long ago. Earthquakes and the landslides they engender hang over Utah's densely populated urban corridor like the sword of Damocles. Ultimately, a disastrous quake is almost certain to come, probably causing large landslides. The scars of several large rockfalls are apparent on the face of the mountains north of Provo.

Though most of the route is among Paleozoic rocks of Utah's great overthrust, north of Oakley the highway is bordered with north-dipping Cretaceous sedimentary rocks that edge the north side of the Uinta Mountains anticline.

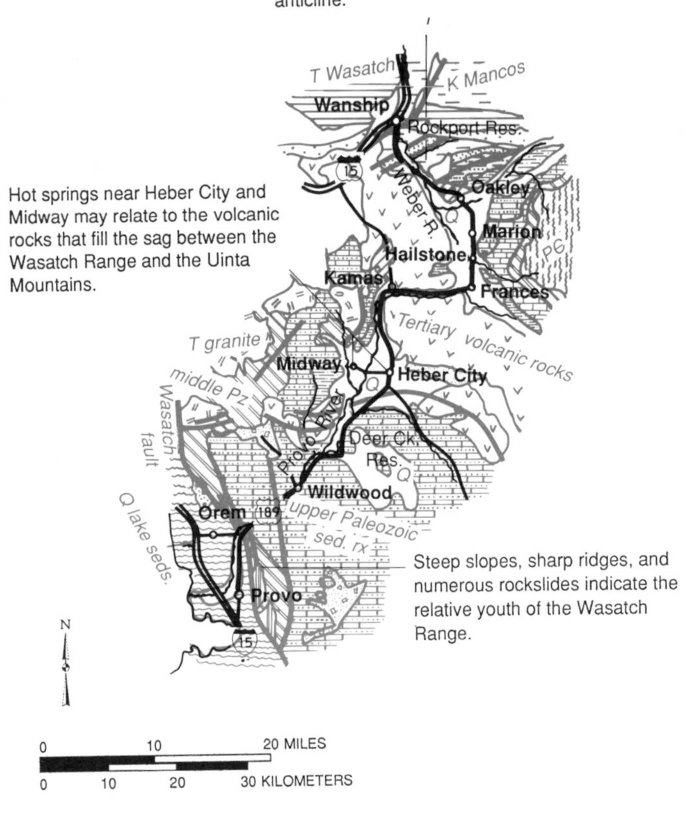

US 189
PROVO — WANSHIP

Thick gravel deposited in the Provo Canyon delta erodes into steep-sided bluffs half concealed in talus.

US 189 travels north on the former delta to the mouth of Provo Canyon. The front of the Wasatch Range here consists entirely of Paleozoic rocks — faulted wedges of lower and middle Paleozoic limestone and shale, and above and behind them a thick sequence of upper Paleozoic limestone, shale, and sandstone, the Oquirrh group, in places as much as 25,000 feet thick. Just think: five miles, measured vertically, of sea-bottom mud, accumulating on the slowly sinking sea floor west of the hingeline! All these rocks then moved east into this area during Cretaceous time as part of the Sevier thrust belt.

Thanks to the Sevier thrust fault, layered limestone and shale of the Oquirrh group, many thousands of feet thick, are in places bent and contorted like sheets of paper.

At Bridal Veil Falls a small stream cascades artistically down across Pennsylvanian / Permian sedimentary rocks of the Oquirrh group.

The Oquirrh group makes up much of the southern end of the Wasatch Range. The strata are bent and broken and in places stood on end by movement of the great thrust of which they and the older Paleozoic rocks are a part, as well as by later movement on the Wasatch fault.

Bumpy pavement near milepost 16 is caused by subsidence and slow downhill creep in these unstable rocks. Rockslides are common here, too, for the strata are finely cut by numerous joints, and break apart easily.

Deer Creek Reservoir gets its water from the Provo River through an intricate system of reservoirs, canals, and tunnels. The water comes from the southern side of the Uinta Mountains, and finally goes to the cities west of the mountains. The big pipeline in the lower part of Provo Canyon carries water from Deer Creek Reservoir to Salt Lake City.

Rocks of the Oquirrh Group extend to Charleston, near the north end of Deer Creek Reservoir. There, they are abruptly cut off by the Charleston thrust fault, one of the major faults of the Sevier thrust belt, which placed Cambrian and late Precambrian rocks on top of Mesozoic rocks. Long after the thrust faulting took place, normal faulting and downward bending formed Heber Valley, which can be thought of as the easternmost expression of Basin and Range faulting.

Heber City lies on the floodplain of the Provo River. North of town the highway follows the river upstream, skirting and then cutting through some Tertiary volcanic rocks, ashflow tuff and breccia, that fill a sort of saddle between the Wasatch Range and the Uinta Mountains. Here we are crossing a pronounced sag in the great anticline that makes up the Uinta Mountains and that continues westward into the Wasatch Range, in both ranges bringing Precambrian rocks to the surface.

East of Hailstone, as the highway follows the Provo River upstream through the volcanic rocks, there is ample chance to look at the light-colored, fragment-filled ashflow tuff and breccia of this sag. The massiveness of the tuff and the many angular fragments in it show that it was deposited from one or more huge, rapidly moving clouds of volcanic ash produced by what must have been a truly massive eruption many times more ferocious than the 1980 explosion of Mt. St. Helens. In places the rock is streaked with whitish and yellowish clay formed as acid volcanic steam decomposed the volcanic rocks. Steam probably continued to rise through the ash long after the eruption.

It seems strange that the Provo River should have cut a canyon through this ridge of volcanic rocks, right on the summit of the anticline. As the Pleistocene Epoch closed, the upper part of the Provo River, coming from the west end of the Uinta Mountains, flowed north across the Kamas Valley to join the Weber River near Oakley. However the lower part of the Provo River, the part we saw in Provo Canyon, gradually eroded headward through the volcanic rocks. It eventually reached the upper Provo, capturing its waters.

North of Kamas the highway leaves the Provo River and almost imperceptibly crosses into the Weber River drainage. This river's canyon also cuts through the Wasatch Range, though in late Cretaceous and early Tertiary time, before the mile-high uplift of Utah and adjoining states, it drained east into Tertiary intermountain lakes. As the Wasatch Range rose, less than 15 million years ago, the Weber, Provo, and Bear rivers managed to deepen their canyons fast enough to keep pace with the uplift. This area figures in the water-supply system serving the urban corridor west of the mountains, where further stream piracy has taken place at the hand of man, who has linked several of the rivers by canals.

At Kamas, we are still on the summit of the sag in the Uinta Mountain anticline. Ski runs visible from Kamas on slopes of the Wasatch Range mark the position of this anticline as it intercepts and becomes part of the Wasatch Range.

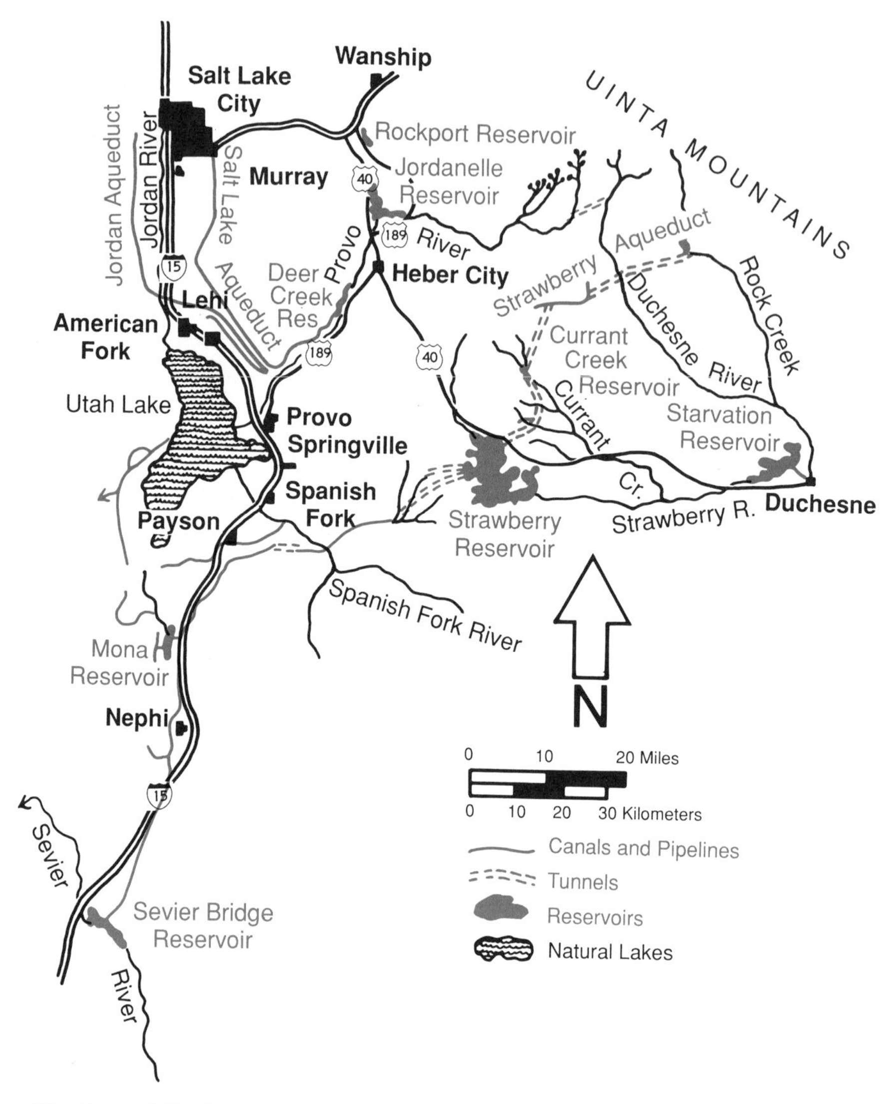

The Central Utah Project reroutes a number of rivers that formerly drained into the Green and Colorado rivers, directing them westward to major population centers.

Between Oakley and Wanship we cross the band of Cretaceous rocks that follows the north slope of the Uinta anticline. We crossed the corresponding band on the south side of the anticline, but didn't see it because it is covered with Paleozoic rocks of the great Sevier thrust belt. Soft gray shales at both ends of Rockport Reservoir, and a band of harder sandstone about halfway along the reservoir, are parts of this belt of Cretaceous rocks.

US 191
Vernal—Wyoming
54 miles/87 km.

North of Vernal, the highway goes through or around successive cuestas and hogbacks of Mesozoic sedimentary rocks tilted up by the rise of the Uinta Mountains. This range consists of one huge east-west-oriented anticline, faulted to some degree along both its northern and southern margins. The hogbacks, erosion-resistant ridges of Mesozoic and Paleozoic rocks that once arched completely across the anticline are separated by valleys eroded in softer rock. Both ridges and valleys run parallel to the mountain range in the vicinity of US 191. However at the western and eastern ends of the mountains they curve around like racetracks and lines of bleachers. The core of the Uinta Mountains, the highest part, consists of Precambrian sedimentary rocks.

The tilted rock units that make up the cuestas and hogbacks and the valleys between them are labeled along this highway, with short notations about the environments in which they were deposited. Driving northward, they read from youngest to oldest, starting with the Cretaceous Mancos shale of the relatively flat landscape around Vernal.

In the valley formed on the Jurassic Morrison formation, the unit that holds the dinosaur remains of Dinosaur National Monument, the highway parallels the hogbacks. Steinaker Lake, an artificial reservoir, uses adjacent hogbacks and the fine, impervious shales of the Morrison formation to contain its waters. Along here the road is bumpy because

North-south section across the Uinta Mountains paralleling US 191

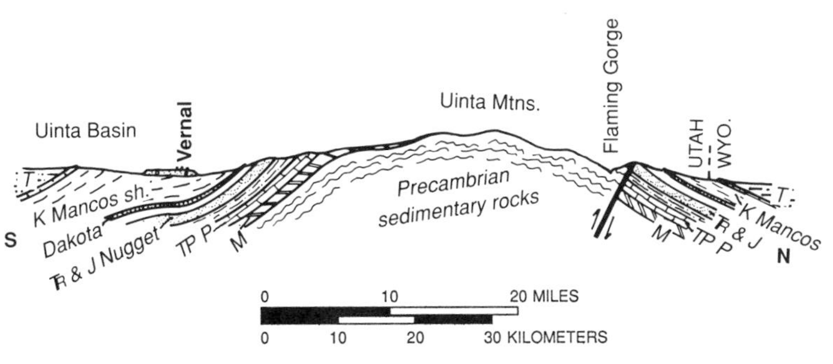

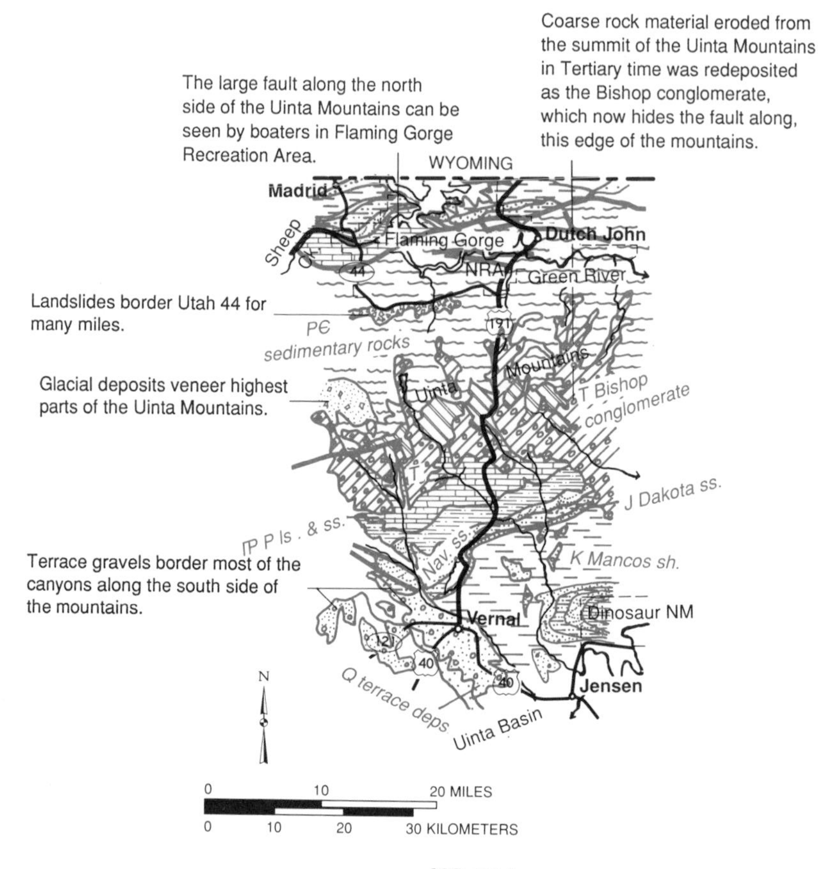

US 191
VERNAL — WYOMING

The open pit phosphate mine near US 191 produces apatite, a phosphate mineral precipitated on the sea floor. Where not diluted by additions of sand and mud, the apatite is a good source of phosphate for fertilizers and other products.

of volcanic ash-derived clay in the Morrison formation, which swells when it becomes wet and shrinks again as it dries.

The large open pit mine on both sides of the highway near milepost 215 produces phosphate from the Park City formation. Between mileposts 215 and 216 the company that operates the mine has erected exhibits that describe the mining and processing of the phosphate, as well as the subsequent reclamation of the land. Phosphate mined and concentrated here is shipped as slurry, a mixture of ground-up concentrate and water, by pipeline to Rock Springs, Wyoming, where it is processed into fertilizers, pharmaceuticals, and other products.

From the same viewpoint you can also see flat-topped mesas that stretch southward from the mountains. These mesas are surfaced with gravel washed from the mountains in Tertiary (probably Oligocene) time, and spread out on surrounding land. Since this conglomerate was deposited, streams from the mountains, no doubt much enhanced during the ice ages, have cut into it, as well as into the surface below it, separating discrete mesas. The fault that edges this side of the Uinta Mountain anticline is hidden below these deposits.

Notable among the Paleozoic rocks along the face of the mountains is the thick, cross-bedded, pink or almost white Weber sandstone, a Pennsylvanian-Permian dune deposit. Because its sand grains are uniform in size and almost round, this sandstone is very porous; it is the reservoir rock for oil produced west of Vernal and at Rangely, Colorado.

Precambrian rocks that make up the main mass of the Uinta Mountains, the core of the range, belong to the Uinta Mountain group (not to be confused with the Uinta formation, a Tertiary unit that surfaces much of the Uinta Basin). They are ancient sedimentary rocks, mostly brick red sandstone and siltstone, about a billion years old. Despite their antiquity, these rocks are only partly metamorphosed, and their sedimentary features are still quite in evidence. This eastern end of the mountains, not as high as the western end, shows few signs of glaciation, though it undoubtedly felt some of the effects of the cold, wet ice ages.

North of milepost 235 the highway descends through the Precambrian rocks to Flaming Gorge Dam and Reservoir. Flaming Gorge was named by John Wesley Powell, leader of the first river float trip (in 1869) down the Green and Colorado rivers and through Grand Canyon, who vividly described "a flaring, brilliant red gorge ... composed of bright vermilion rocks." It is the Precambrian sedimentary rocks that give the gorge its red color. They are well exposed near the dam: red quartzite interlayered with red shale and siltstone, sediments resembling those known to be deposited today on deltas and river floodplains. There are 20,000-25,000 thousand feet of these sedimentary rocks here — four to five vertical miles!

North of the dam, at milepost 247, the Precambrian rocks are steeply dipping, almost vertical. The cliff to the north marks the major east-west reverse fault that edges the north side of the mountain range. Though topographically higher, the cliff is actually the down side of the fault, and consists of the Nugget sandstone, a Jurassic dune sandstone equivalent to the Navajo sandstone of Plateau country Utah.

North of the fault we come into younger and younger units: first a racetrack valley of Jurassic shale, the Morrison formation again; then the Dakota sandstone hogback; then a wider valley in soft Cretaceous Mancos shale, where the highway is bumpy because of swelling and shrinking clays. To the west, Flaming Gorge Reservoir widens, spreads out, in the area of the soft Mancos shale. The hogbacks north of milepost 252, near the Wyoming line, are capped with Cretaceous Mesaverde sandstones, complete with seams of coal. Notice that the strata dip less and less steeply farther away from the Uinta Mountains.

Utah 44, an alternate route to Wyoming, stays south and west of Flaming Gorge Reservoir. It crosses Sheep Creek, where a side road leads into Sheep Creek Canyon Geological Area. Rocks along the route are labeled, as is the site of a disastrous flood.

A northwestern arm of Flaming Gorge Reservoir points to Sheep Creek Canyon. Cliffs on the left are upward-faulted Nugget sandstone, equivalent to the Navajo sandstone farther south. The valley and slope to the right are surfaced with Permian to Triassic rocks.

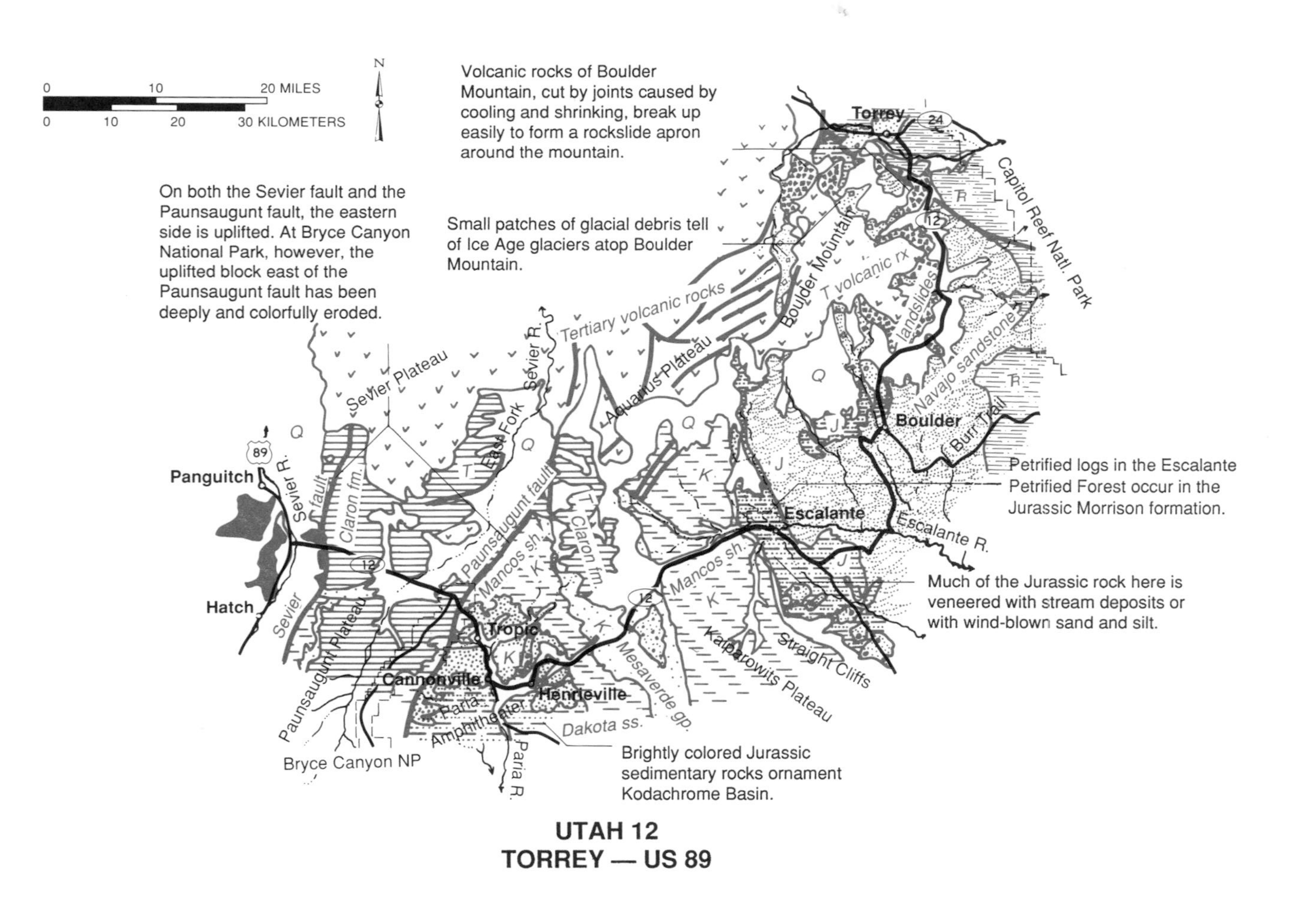

UTAH 12
TORREY — US 89

A rounded knob of Navajo sandstone rises above surrounding pines.

Utah 12
Torrey—US 89
near Panguitch
121 miles/195 km.

As it leaves Utah 24 near Torrey, this route heads south through red-brown Triassic mudstone and siltstone, some of it decoratively marked with 230-million-year-old ripple marks. A short distance to the south the Jurassic Navajo sandstone is faulted upward above these red rocks, and appears as sculptured hills projecting from the trees. Patterns of cross-bedding in the Navajo sandstone intersect evenly spaced joints; rills eroded along the joints give this rock a checkerboard appearance. The highway runs southeast along the fault between the two formations after crossing Fremont Creek.

When it leaves the fault the road climbs onto landslide debris that forms most of the eastern shoulder of Boulder Mountain, a large plateau built up of many successive Tertiary lava flows, layers of breccia, and quantities of volcanic ash. Younger lava flows cap the north end of the mountain.

Near Rock Creek the highway follows the contact between the Navajo sandstone and the landslide slope. The immense pile of volcanic rock of Boulder Mountain, possibly sagging into its own partly emptied magma chamber, seems to have caused the rock layers below it to sag, too, so that surrounding sedimentary rocks dip toward the mountain. Basalt boulders near Rock Creek are part of a landslide. Roadcuts show patches of very light-colored volcanic tuff, ash from the Miocene volcano.

The tuff near milepost 113 is hard and shaly, firmly welded into rock that looks almost like porcelain. This is ashflow tuff formed as an incandescent ash cloud shot down the flank of the volcano, welding itself together when it came to rest. Ashfall tuff, which cools before reaching the ground, is generally softer.

From many points along this route superb views look east to the Henry Mountains, and beyond them, 100 miles away, the La Sal Mountains. Both of these ranges formed in Miocene time when molten magma pushed up through sedimentary strata, doming up the uppermost layers. The sedimentary layers have now eroded from them: The peaks of the ranges are the igneous rock of the intrusions.

Many of the flat-topped mesas that characterize Plateau country scenery can be seen from highway vantage points. In the wild and beautiful terrain between the highway and the Henry Mountains, flat-lying rocks swoop up into the ridges of Waterpocket Fold, marking the position of Capitol Reef National Park. Prominent light-colored rock exposed along the summit of this fold is the Navajo sandstone, which erodes into the domes that give Capitol Reef its name. Just

The Henry Mountains rise beyond the long ridge of Waterpocket Fold, in Capitol Reef National Park. Foreground boulder-strewn gravel is part of a landslide.

Bent trunks of an aspen grove indicate downslope soil movement or creep when the trees were young.

about all the sedimentary rocks you can see from the highway are Mesozoic.

The highway travels for many more miles on the landslide apron that surrounds Boulder Mountain. The hummocky surface, irregularly spotted with marshy depressions, is a characteristic feature of landslides. Probably the slides were most active in Pleistocene time, when an ice cap topped Boulder Mountain, but there is plenty of evidence of recent movement, including bent tree trunks and damage to the pavement. Soil creep, a much slower movement, helps to round the hills and slopes.

Elevation at the highway summit at milepost 101 is 9400 feet. Beyond this point Utah 12 travels in a more southwesterly direction. The viewpoint a short distance beyond the summit overlooks the surrealistic landscape of the Escalante country, surfaced with Navajo sandstone and deeply furrowed by the Escalante River and its tributaries. More Mesozoic rocks. In the distance to the south are the Straight Cliffs, carved in gray Cretaceous rocks, edging the Kaiparowits Plateau. The round dome of Navajo Mountain, a single laccolith still covered with Jurassic rocks, is 70 miles to the south. These and other features are identified at the viewpoint near milepost 99.

Southwest of the viewpoint the highway descends rapidly to the town of Boulder, with ponds and low marshy areas, as well as crooked tree trunks, to show that landslides exist here, too. A side road to

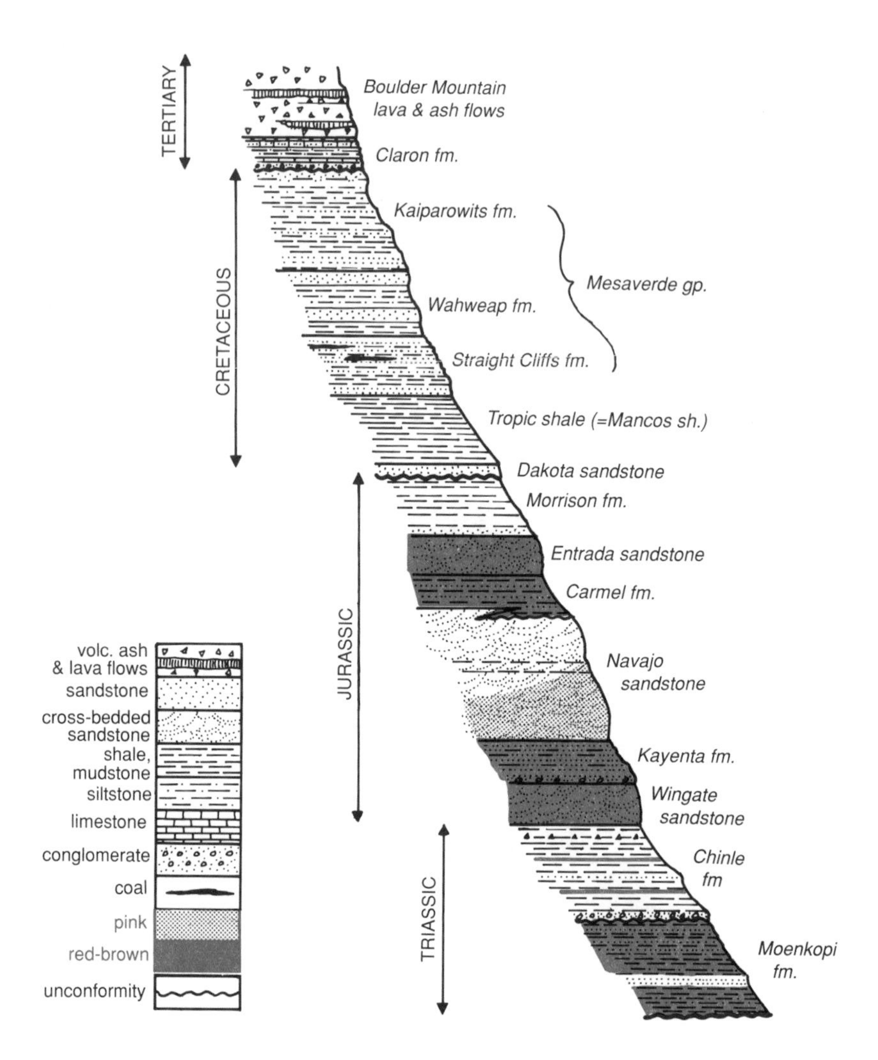

Stratigraphic diagram of rocks visible from Utah 12

Garkane Power Plant reminds us that we're driving through coal-bearing Cretaceous rocks.

The town of Boulder, with springs to furnish its water, seems an oasis in a desert. Beyond the town, the Navajo sandstone is capped by thin, hard rimrock—the Jurassic Carmel formation. The road travels along a knife-edged ridge worthy of a good thriller, with nearly 1000 feet of drop-off in cliffs to either side. There is plenty of opportunity here to look closely at the many interesting features of the Navajo sandstone: its large-scale, sand dune-style cross-bedding interrupted by thin, flat, pale pink layers of siltstone that accumulated on flat

A foreground of hummocky landslides frames rugged country etched by the Escalante River and its tributaries. Bare white rock is the Navajo sandstone.

areas between the Jurassic dunes, occasional bands of silty limestone that show that some interdune flats held shallow ponds that came and went with shifting of the dunes. Large, arching recesses are typical of the Navajo sandstone and several other southwestern dune sandstones. The formation extends from Wyoming, where it is called the Nugget sandstone, to the southern tip of Nevada, and covers an area about as large as today's Sahara Desert. In Jurassic time this area lay at the same latitude as today's Sahara, a latitude characterized by dry easterly winds that suck up moisture but bring little or no rain.

Outcrops of Navajo sandstone near Boulder show joints that intersect at odd angles, so that weathered surfaces appear like biscuits jammed together in a pan.

The rugged Escalante country, surfaced with Navajo sandstone, is trenched by tributaries of the Escalante River. The floors of some canyons expose colorful Triassic shales.

The highway descends to and crosses the Escalante River, quite small here despite its deep canyon, and then climbs up through the Navajo sandstone again. Here the sandstone is partly pink, as it is farther west at Zion. The color — attributable to tiny amounts of iron oxide or hematite — doesn't follow the stratification, however, or for that matter the cross-bedding. It seems to be a later addition, with oxidized iron added by groundwater moving slowly through the porous sandstone.

Large-scale cross-bedding of the Navajo sandstone, coupled with its fine, rounded, frosted grains, tell us it was deposited on the lee faces of ancient dunes. Horizontal layers are interdune deposits.

Cretaceous rocks form the Straight Cliffs, part of the Gray Cliffs of Utah's Grand Staircase. Above them is the Kaiparowits Plateau.

Rising out of a canyon near milepost 69, we see the lava-capped Aquarius Plateau, its pink cliffs smaller editions of those at Bryce Canyon National Park. From here we can also see the north end of the Straight Cliffs and the Kaiparowits Plateau behind them.

Formations are named for nearby geographic features. Geologists studying a rock unit measure and describe a type section, a standard against which other exposures of the same unit can then be compared. The three units that make up the Straight Cliffs — the Tropic shale, Straight Cliffs formation, and Kaiparowits formation — all take names from sites near this highway. These formations are equivalents of the Mancos shale and Mesaverde group, which get their names from places in Colorado.

At Escalante, we cross the Escalante River again. Did you think it was on the other side of the mountain? It was — and it cuts right through the plateau we have just crossed. The entrance to its steep-walled canyon shows in the cliffs east of town. The river maintained an older course, developed on a much flatter surface, when it carved deeply into the rock layers in response to regional uplift and lowering of base level as the Colorado River cut downward.

Near Escalante the Navajo sandstone, along with units above and below it, curves upward as part of a monocline defining the north end of the Kaiparowits Plateau. One of the rock units brought to the surface along this monocline is the Morrison formation, a soft, colorful, easily eroded shale that contains a high proportion of volcanic ash. In the Escalante Petrified Forest a few miles west of Escalante, the Morrison formation contains tree trunks and stumps preserved, like

A silicified tree trunk lies on the ground in the Escalante Petrified Forest. Collecting fossil wood is prohibited.

those in Arizona's Petrified Forest, by silica deposited in the wood by groundwater. The trees may well have been downed by a volcanic explosion similar to that at Mt. St. Helens in 1980, and later impregnated with silica derived from the fine volcanic ash. Red and yellow jasper color the petrified wood.

Farther west, the highway goes through gray shale and yellowish sandstone of the Straight Cliffs and Wahweap formations, which we have already seen at a distance. Cretaceous rocks range in color from grayish yellow to gray, with none of the white or red tints seen in Triassic, Jurassic, and Tertiary rocks. Reds and pinks come from tiny amounts of oxidized iron. The Cretaceous rocks contain iron minerals as well, but they also contain enough plant material to keep the iron from oxidizing.

Near the highway a tributary of the Escalante River displays a braided channel characteristic of southwestern watercourses, where on-again off-again streams swing from side to side as they choke themselves with too much sand. Tamarisk and Russian olive trees, not native to this area, have invaded the stream banks; both of these prodigious water users are now common along western rivers.

From milepost 39 we get our first views of Bryce Canyon, with its ornately sculptured pink Tertiary (Paleocene) lake deposits — silty limestone and limy siltstone — carved in the rim of the Paunsaugunt Plateau. Between mileposts 34 and 33, the mesa ahead tells us a little about the history of the Paria Amphitheater and the sculptured walls of Bryce. Though rock strata of the mesa dip northward, the mesa top slopes gently to the south. Its beveled top, thickly covered with gravel, is an old erosion surface created before the present Paria Amphitheater. When the mesa surface formed the amphitheater floor, Bryce's eroded scarp may not have been as high as it is now.

That scarp developed along the Paunsaugunt fault, although it has eroded back from it now. Oddly, we are on the upthrown side of the fault, and Bryce's scarp, well above us, is on the downthrown side — a feature illustrated in the discussion of Bryce Canyon National Park in Chapter IV.

East of Henrieville and bordering the highway from there to Cannonville, prominent pink- and white-striped Jurassic rock is eroded into hoodoos. Watch for small faults offsetting the pink and white layers. Cannonville nestles at the foot of cliffs of this sandstone. Kodachrome Basin south of Cannonville, hardly more colorful than the country we have come through, gets its name from these same peppermint-striped rocks.

The road crosses the Paria River at milepost 26, and follows it upstream through mustard-colored Cretaceous rocks marked with thin seams of coal. The town of Tropic is the type locality for the Tropic shale, one of the coal-bearing units. West of Tropic, Cretaceous rocks erode in gentle slopes and steep, irregular ridges. The Paunsaugunt fault, along the southeast edge of the Paunsaugunt Plateau, shows up north of the highway just beyond the Bryce Canyon National Park

Eroded spurs of the Aquarius Plateau expose Tertiary lakebeds similar to those at Bryce.

boundary sign. There the Cretaceous rocks abut much younger, pinker limestone and siltstone of the Claron formation. Because of ravines eroded across it, the contact appears to zigzag; drive a little farther to see that the fault is really nearly vertical.

West of the fault the highway winds among some of the sculptured pink rock. Erosion of Tertiary lake sediments is particularly rapid here and within the national park, partly because they are soft, partly because relief between the plateau surface and the Paria Amphitheater is great. The high surface of the plateau is protected by a resistant dolomite layer at the top of the Claron formation.

The rest area east of milepost 10 provides views west and northwest across the Panguitch Valley to high volcanic mountains of the Markagunt Plateau. A little farther west the highway drops into Red Canyon, with the Claron formation, redder than at Bryce, carved into knobby pinnacles. The Sevier fault, the western boundary of the Paunsaugunt Plateau, is just west of milepost 3, and appears north of the highway, where dark gray lava flows abut Claron formation lakebeds. Crossing the fault, the highway descends a long alluvial fan toward the Panguitch River and highway US 89.

Utah 14
Cedar City — Long Valley Junction
41 miles/66 km.

Traveling east, this highway goes up Cedar Canyon, one of the few breaks in the great escarpment of the Hurricane Cliffs, which separate the Colorado Plateau from the Basin and Range deserts to the west. Close to the mouth of the canyon, the route crosses the Hurricane fault, responsible for the Hurricane Cliffs. The fault is not a clear-cut single break, but a two-mile-wide fault zone, a complex of many faults and small anticlines and synclines. As you can tell, movement on this fault lifted the east side several thousand feet above the west side. The steep dip of the Triassic sedimentary rocks, as well as of lighter-colored Jurassic rocks, results from drag along the fault.

The Triassic rocks, mostly deep red-brown to purple siltstone and mudstone, were originally deposited in a floodplain-delta environment close to the shore of an ancient sea. In color, grain size, and layering or bedding they contrast with the pale Jurassic Navajo sandstone, whose sweeping cross-bedding and fine, even, rounded grains show that it accumulated as sand dunes on an ancient desert. This is the rock that forms the massive cliffs of Zion, farther south, but here it is highly broken by faulting. Pink and cream-colored, candy-striped sandstone and siltstone visible on up the canyon are Jurassic also.

Section parallel to Utah 14 between Cedar City and Long Valley Junction.

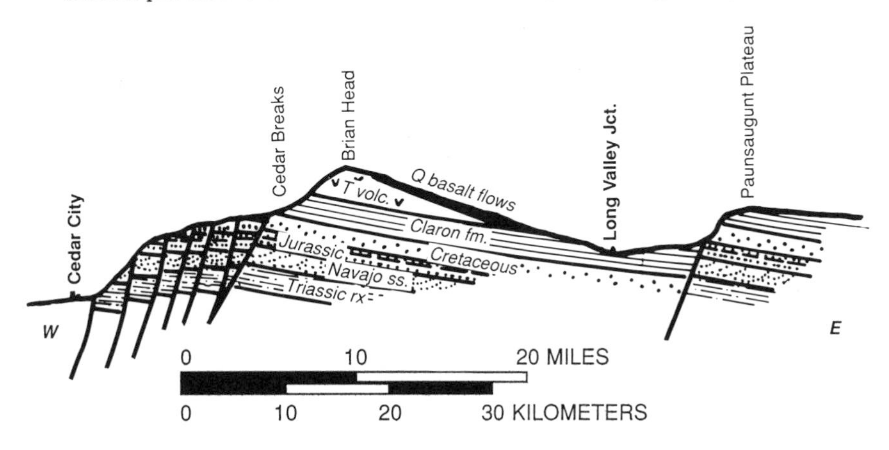

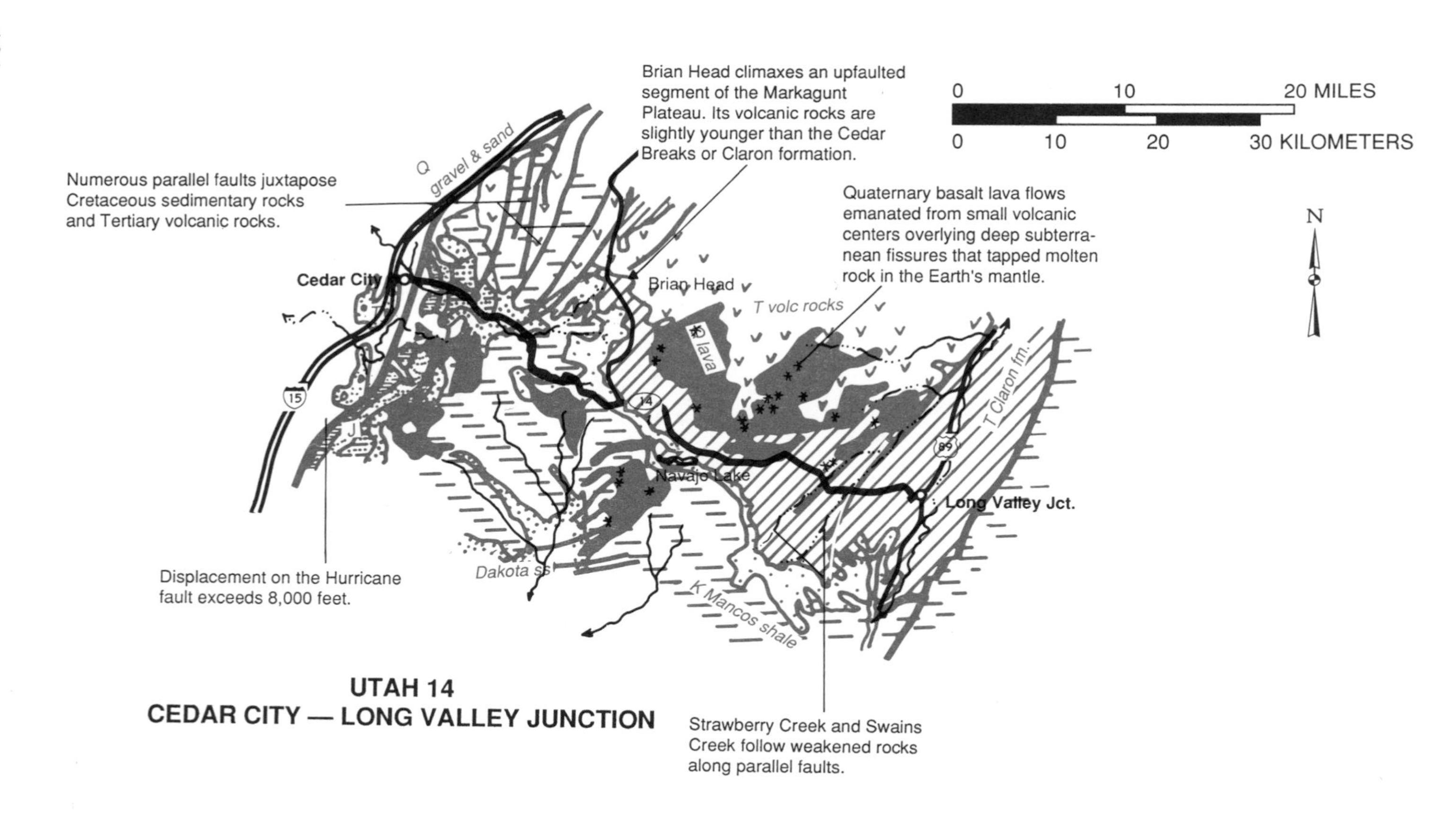

UTAH 14
CEDAR CITY — LONG VALLEY JUNCTION

Cretaceous rocks come into the picture still farther up the canyon, near milepost 8. By and large they are grayish and yellowish gray, less colorful than the Triassic and Jurassic rocks. And less steeply dipping — they level out at the top of the Hurricane Cliffs.

In all of southern and eastern Utah, Cretaceous rocks consist of three basic units: the thin but resistant Dakota sandstone at the base; soft, easily eroded, dark gray marine shale, the Mancos or Tropic shale, just above it; and a thick sequence of sandstone, shale, and coal, the Straight Cliffs and Wahweap and Kaiparowits formations, at the top. Farther east the upper sandstone-shale-coal units are called the Mesaverde group.

Taken together, these rocks tell of the advance and retreat of the Cretaceous sea — a fairly shallow sea that swept over this area from the east. The Dakota sandstone is the shoreline deposit — beach and near-shore sandstone. The dark shale above it — here called the Tropic shale — was deposited in the sea. But because the Cretaceous sea came from the east, it was not here as long as it was in eastern Utah, and the shale unit is not as thick as its equivalents there. The Straight Cliffs, Wahweap, and Kaiparowits formations represent the fluctuating retreat of the sea, with repetitions of shoreline sandstones alternating with mudstones and coal that accumulated in lagoons along a low-lying coast. Mostly shale and sandstone, the Kaiparowits formation includes several layers of limestone made almost entirely of shells and shell fragments; look for them between mileposts 13 and 14.

Ahead of and above the highway, particularly near the campground at milepost 12, we get occasional glimpses of the breaks above — scenic, ornately eroded patches of Tertiary limestone and siltstone, among them Cedar Breaks in Cedar Breaks National Monument. Carved in soft Tertiary lake deposits of the Claron formation, several of these breaks ornament the west edge of the Markagunt Plateau. In this area the silty, limy lake deposits are redder than their equivalents at Bryce.

From the viewpoint near milepost 19, the panorama to the south includes the northern part of Zion National Park. The viewpoint and the highway are on the dividing line between the Markagunt Plateau to the north and the Kolob Terrace to the south. The Kolob Terrace is surfaced with the same Cretaceous rocks we've been seeing, mostly tree-covered. Far to the south are knobby ridges of Navajo sandstone. The Markagunt Plateau north of us is surfaced with the Claron lake deposits and in places capped with Tertiary and Quaternary volcanic rocks.

The breaks that characterize the west rim of the Markagunt Plateau are eroded in soft lakebeds of the Claron formation.

A short distance east of this viewpoint is the road to Cedar Breaks National Monument (discussed in Chapter IV), where these rocks are fully exposed. The lake in which they formed was one of several freshwater lakes that in early Tertiary time developed in wide basins between the rising ranges of the Rocky Mountains.

Continuing eastward, Utah 14 skirts some of the young volcanic rocks that cap much of the Markagunt Plateau. East of Navajo Lake, a manmade reservoir, the highway travels among some of the lava flows, remarkable for their rough, new-looking basalt. They came from numerous small volcanic centers, some of them shown by asterisks on the map. They completely disrupted preexisting drainage patterns, damming many small ponds now partly ingrown with cattails and other moisture-loving plants. Such disrupted drainage is further evidence of the youth of these lava flows. As ponds fill in with plant matter and silt, and as lava flows erode, through drainage will eventually be reestablished.

The Tertiary volcanic rocks that underlie the basalt flows came, by contrast, from large composite or stratovolcanoes. The Tertiary lavas were silicic, thicker and more viscous than runny Quaternary basalt lavas. Because of this, Tertiary lava flows alternate with volcanic ash and breccia created by explosive eruptions. Ash, flows, and breccia built up into towering mountains, many of which were later blasted away in renewed explosions.

The Markagunt Plateau slopes eastward, crossing two faults near Strawberry Creek and Swains Creek. The highway proceeds down this eastward slope to Long Valley Junction and US 89. This junction is on the drainage divide between the East Fork of the Virgin River, flowing south, and the Sevier River, flowing north. East of Long Valley Junction, the Paunsaugunt Plateau rises along the Sevier fault, which brings Cretaceous rocks abruptly to the surface. Back from its rimlike edge, the Paunsaugunt Plateau is capped with the same Claron formation lake deposits that we've seen on the west and south edges of the Markagunt Plateau. For more about the Paunsaugunt Plateau, see Bryce Canyon National Park in Chapter IV.

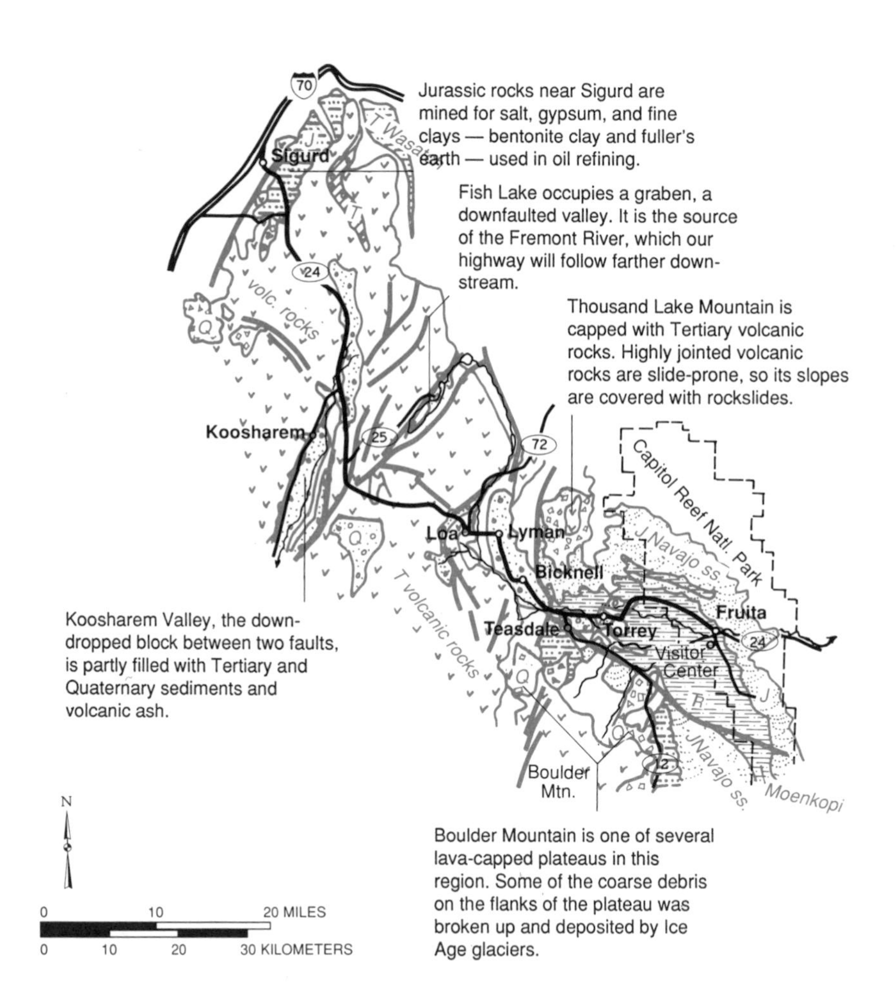

UTAH 24
SIGURD — CAPITOL REEF

Gypsum plants near Sigurd produce plaster, plasterboard, and other building products from gypsum deposited in an evaporating Jurassic sea.

Utah 24
Sigurd—Capitol Reef
72 miles/116 km.

Sigurd is the site of two large gypsum plants. The gypsum is mined from the Arapien Shale, the Jurassic rock unit that forms all the light-colored hills east of the town, and made into plasterboard and other products. Behind the mills are heaps of spent shale from which gypsum has been removed.

Though it is a sedimentary rock, the Arapien shale displays none of the regular banding characteristic of most sedimentary rocks. This is because gypsum, formed originally by evaporation of sea water, tends to flow, pushing upward wherever pressure is lower than elsewhere. Gradual movement of the gypsum, and probably of associated salt, disrupts layering in both gypsum and overlying rocks, and is responsible for the thick, bulging appearance of the hills here.

Gypsum mines are south and north of the highway near milepost 13. Look for chunks of almost pure gypsum along the highway. Softer than most other minerals, gypsum can be carved with an ordinary pocket knife; pure, dense gypsum, sometimes tinted pink, is known as alabaster.

East of the Arapien shale hills the route enters a volcanic area of lava flows, breccia, and light-colored volcanic ash. Most of the rock is breccia—broken fragments of volcanic rocks mixed with volcanic ash. This massive rock forms much of the north end of the Sevier Plateau, which the highway crosses here.

Unusually deep gullying in fine river-deposited sediments of the valley floor near milepost 18 is probably due to introduction of grazing animals.

Crossing the Sevier Plateau, we get good views of Fish Lake Plateau to the east and Awapa Plateau to the southeast. These uplands are surfaced with volcanic rocks like those we have just seen. East of milepost 24 the highway winds down into flat-floored Plateau Valley, a long north-south graben formed by subsidence of a sliver of land between two parallel faults. East of the junction with Utah 62, our route climbs the east side of this valley, crossing one of the faults. From this part of the highway we look down on the town of Koosharem and along the straight fault-edged graben south of it. Both the Sevier Plateau west of here and the Awapa Plateau to the east are covered with thick layers of Tertiary volcanic rocks—breccia, lava flows, and tuff produced by a composite or stratovolcano. The valley is edged with

Tertiary volcanic rocks include cliff-forming breccia, made of broken volcanic rock, overlying a layer of mudflow deposits. Both are products of explosive eruptions.

broad alluvial fans made of materials whittled from adjacent volcanic plateaus.

The highway turns east across the divide between the Awapa Plateau and the Fishlake Plateau to the north. Much of the geology here is hidden by soil and sagebrush. From milepost 41, Thousand Lake Mountain is directly ahead. Though it is built of Miocene volcanic rock capped with Pliocene lava flows, its slopes are almost entirely landslides. The slides conceal much of the geology, but here and there Tertiary lakebeds show their pink color.

Farther east are the pointed summits of the Henry Mountains, rising above the Plateau country of eastern Utah. These peaks are a cluster of small igneous intrusions that in Oligocene time punched through Paleozoic and Mesozoic sedimentary rocks, then spread beneath the Cretaceous layers above them. Cretaceous rocks have now eroded from the crests of the intrusions, baring the granitic igneous rock.

To the south is Boulder Mountain, a Miocene shield volcano. Between it and Thousand Lake Mountain to the north, some of the red sedimentary rocks of the Plateau country stand out, contrasting in color with the gray volcanic rocks of the mountains.

Tributaries of the Fremont River have carved deep canyons into the side of the Awapa Plateau, visible as the highway approaches Loa. Near Teasdale they join the Fremont River, which drains the Fish Lake Plateau and Thousand Lake Mountain.

Loa, Lyman, and Bicknell lie on broad gravel-covered slopes below these mountains. Most of the gravel was deposited in Pleistocene time when streams from the mountains were heavily loaded with coarse glacial debris. Between Lyman and Bicknell gray lava boulders litter the surface, some of them coated with white caliche deposited by carbonate-rich groundwater. The highway remains on the Pleistocene deposits to and beyond Bicknell, crossing the Fremont River near Bicknell. There are some lava flows north and east of Lyman.

At Bicknell, watch for a visible and quite sudden change from volcanic rocks to sedimentary rocks. A major north-south fault between Bicknell and Teasdale brings Mesozoic sedimentary strata, the rocks that characterize the Plateau country of southeastern Utah, to the surface. Like those described above, this fault is largely concealed by landslides. But its topographic results are clear: North of Thousand Lake Mountain and the Fish Lake Plateau, the fault defines the east edge of the Wasatch Plateau; to the south, it separates Boulder Mountain from the Awapa Plateau. Just north of Bicknell you can recognize the Arapien shale, the rock unit we saw on the other side of the mountains, with distorted bedding and patches of gray or white gypsum.

Red cliffs east of Bicknell are the Wingate sandstone, a Jurassic dune deposit, tilted steeply along the fault. Massive, cross-bedded, knobby white and pink rock is the Navajo sandstone, another Jurassic dune deposit.

Triassic rocks appear in the canyon of the Fremont River. The dark brick red ones are mudstone and siltstone of the Moenkopi formation, thin layers deposited in a mudflat or tidal flat environment. The formation is capped with the resistant Shinarump conglomerate, the lowest layer of the Chinle formation.

Nestled between Boulder Mountain to the southwest and the uplands around Capitol Reef National Park, Torrey lies on a pediment cut in rocks of the Moenkopi formation. Lava boulders and other coarse, glacially quarried, stream-deposited debris from Thousand Lake Mountain and the Fishlake Plateau cover parts of the surface.

Utah 12 branches off to the south beyond Torrey — a particularly scenic and geologically interesting route to Bryce Canyon. Near the junction are more exposures of the Shinarump conglomerate and the softer Moenkopi formation redbeds below it.

The Shinarump conglomerate, a remarkable rock unit spread thinly over much of northern Arizona and southern Utah, is mostly sandstone here. Elsewhere it contains pebbles derived from Triassic mountains of central Arizona. It seems unbelievable that such a sheet of gravel and sand could develop over such a wide area. It is about 30 feet thick here but thins markedly near the Capitol Reef National

Ornately sculptured cliffs of red Jurassic siltstone border the Fremont River. Note the talus slopes below the cliffs. The long slope angling down from the left is a remnant of an older talus slope.

Park Visitor Center. It is the lowest part of the Chinle formation, other parts of which are weak, easily eroded mudstone full of volcanic ash, prone to erode into colorful badlands. Above the Chinle formation is a Jurassic rock unit: the thick, cliff-forming Wingate sandstone.

White ledges of Permian limestone appear in the gorge of Sulphur Creek and some of its tributary canyons, south of the highway.

At Panorama Point, enjoy the good views of Capitol Reef and the Henry Mountains. Capitol Reef is part of a 70-mile-long monocline known as Waterpocket Fold — a one-way fold, bent down to the east. Here the word "reef" is used in its older sense, meaning a barrier. It is certainly that. Capitol Reef National Park is discussed in Chapter IV. The park visitor center is constructed of Moenkopi sandstone, with many slabs displaying ripple marks, raindrop pits, and mudcracks formed in Triassic time. On the hill above the visitor center, these sloping rocks are beveled by erosion, then covered with gravel and lava boulders washed from the Fish Lake Plateau in Pleistocene time.

UTAH 150
KAMAS — WYOMING

Utah 150
Kamas—Wyoming:
The Western Uintas
56 miles/90 km.

East of Kamas, Utah 150 follows Beaver Creek into the heart of the Uinta Mountains. A horseshoe-shaped arc of Paleozoic rocks, bent upward as part of the Uinta Mountain anticline but now eroded away from the central part of the mountains, surrounds this western end of the Uintas. Chippy gray Pennsylvanian limestone, as well as more massive darker gray limestone, appears at roadside level and here and there on slopes above the highway.

On up Beaver Creek Canyon are older rocks. Mississippian strata appear between mileposts 2 and 3, and include a massive, fossil-bearing limestone that is part of a thick sheet of limestone that extends from Grand Canyon to Wisconsin and Michigan — evidence of a sea that spread over much of the continent. In this immediate area the Mississippian limestone lies right on Precambrian sedimentary rocks; their contact comes between mileposts 6 and 7, above Beaver Creek Narrows. No Devonian, Silurian, Ordovician, or Cambrian rocks exist here. They may have existed at one time, but were eroded off before the Mississippian limestone was deposited. Small remnants of Cambrian strata do appear in some parts of the horseshoe that encircles this end of the Uintas.

At the upper end of Beaver Creek Narrows the valley widens out, marking the lower limit of Pleistocene glaciation. The highway cuts through a glacial moraine near Fall Creek — loose rocks drop from it onto the highway. Others near Yellow Pine and Beaver Creek campgrounds document the gradual retreat of the glacier that occupied

Glacial deposits are identified by their angular boulders and completely unsorted mix of boulders, pebbles, sand, silt, and clay. Some of the boulders may be grooved from grinding against other rocks.

this valley. Moraines can be recognized as bumpy hills curving across stream valleys, and as unsorted, unstratified mixes of gravel with angular rock fragments not rounded by stream action.

More glacial debris, in some places mixed with river-deposited gravel, is visible near milepost 12 where the highway crosses the Provo River. From here to Bald Mountain Pass the highway follows this river.

Precambrian sedimentary rocks and glacial debris derived from them become more and more common as we go farther into the

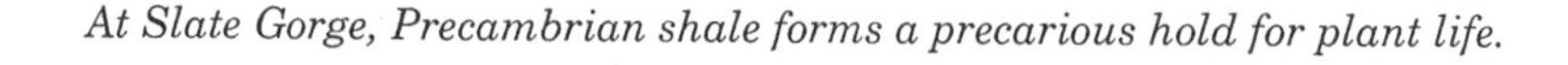

At Slate Gorge, Precambrian shale forms a precarious hold for plant life.

mountains. The Precambrian strata here, called the Uinta Mountain group, seem to have been deposited in a narrow fault valley or rift that about a billion years ago cut deeply into the North American continent. The rocks include dark gray, reddish, and black shale and sandstone, all well displayed at Slate Gorge. The rock is not slate, which breaks across the bedding rather than parallel to it, but shale, which breaks along bedding planes. Glacial boulders include white Tintic quartzite from the high central part of the range.

South of the highway, in cliffs and steep slopes of Precambrian rock, rockslides are prominent landscape features. Many of them continue to move, shifting perhaps a few inches or a few feet a year, preventing soil development so that vegetation can't establish itself.

The Provo River's headwaters are at Lilly Lake, Lost Lake, and other small lakes to the west and north, in the same general area as the headwaters of the Weber, Bear, and Duchesne rivers, making this area an important water source for the state of Utah.

Upper Provo River Falls cascade across thinly bedded Precambrian sandstone.

Evenly bedded limestone near Bald Mountain Pass belongs to the billion-year-old Uinta Mountain group.

Bald Mountain (11,949 feet), ahead as the highway leaves Lilly Lake, shows well its horizontal layers of Precambrian rocks of the Uinta Mountain group: green-gray shale and orange quartzite. The mountain rises above the high point in the Uinta Mountain arch, the great faulted anticline that makes up the range. Above timberline, between 10,000 and 11,000 feet at this latitude, the rocks of which Bald Mountain is composed are clearly exposed.

Precambrian limestone interlaced with the trails of ancient marine animals have been called spaghetti beds.

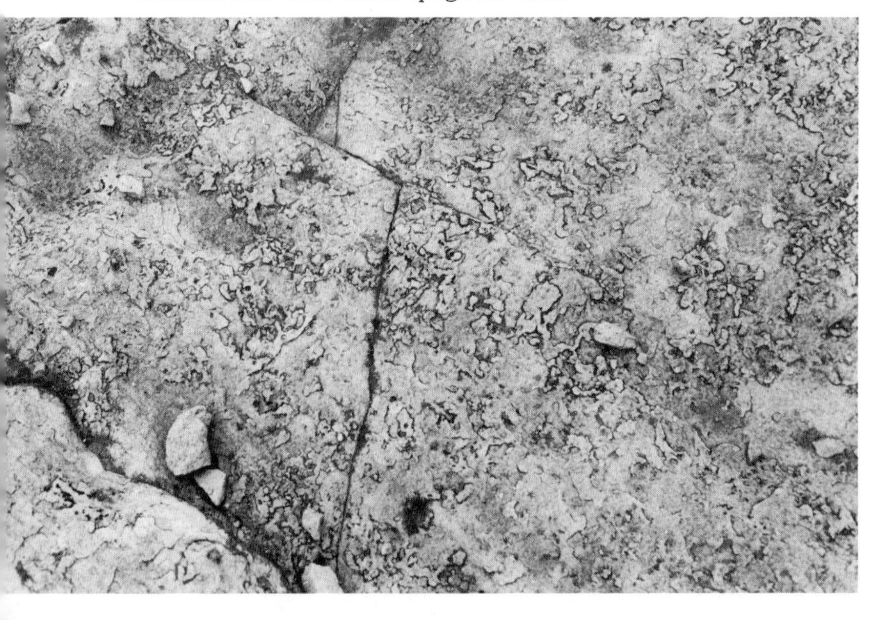

Near Bald Mountain Pass are many exposures of an evenly bedded light gray limestone, each layer two to three inches thick. This rock contains some of the oldest fossils known, fossil algae classed under the catch-all name stromatolites, which develop as small domes made up of many more or less concentric layers of calcium carbonate. Here the stromatolites appear as bumps on flat rock surfaces. Near the Bald Mountain trailhead similar rocks contain what appear to be the twisting trails of animals that burrowed through the limy muds of the Precambrian sea floor more than a billion years ago.

At the viewpoint near milepost 30, Hayden Peak and Mt. Agassiz, both over 12,000 feet in elevation, rise to the northeast and east. Mt. Agassiz is named after Louis Agassiz, a Swiss biologist-paleontologist-geologist who in 1836-1840 studied the movements and effects of glaciers in the Alps, demonstrating decisively that much of Europe had once been buried in glacial ice. Later, at Harvard University, he revolutionized the teaching of natural history, encouraging his students to observe things for themselves, to "strive to interpret what really exists." Saddles north and south of Hayden Peak mark the positions of some of the faults that run along the crest of the Uinta Mountains.

The view from here also gives a good picture of the once ice-capped plateau of the western Uintas, where at the close of the ice ages,

Hayden Peak, seen here across Moose Horn Lake, was named for Ferdinand V. Hayden, a geologist who between 1867 and 1879 led a government-sponsored geologic and geographic survey of the western territories.

A quartzite boulder stranded by a melting glacier now offers protection to young trees.

glacial ice stagnated and melted, leaving an abundance of lakes and a widespread blanket of rocky glacial ground moraine. The area into which the highway now descends, between Bald Mountain Pass and Hayden Pass, displays these features well. The numerous white quartzite boulders, transported by glaciers from some of the high peaks, come from part of the Uinta Mountain group, and are one of the most resistant of the Precambrian metamorphosed sedimentary rocks.

North of Hayden Pass, the highway descends the Hayden Fork of the Bear River, which flows in a gentle S-curve northward to the Wyoming border. For much of its route the valley displays the broadly gouged cross section typical of a glaciated valley. Recessional moraines, ridges of unsorted sand and rock that accumulated at the successive tips of a retreating glacier, curve across it at several points. The river's route probably was established in pre-ice age times; glaciers smoothed and straightened it.

Near Beaver View and Hayden Fork campgrounds, we once more cross the horseshoe-shaped band of Paleozoic sedimentary strata that surrounds the west end of the Uinta Mountains. Here these rocks are pretty well concealed by glacial debris. The fault that defines the north edge of the Uinta Mountain uplift crosses beneath the highway between Hayden Fork and Stillwater campgrounds. North of it is a large intermountain basin filled with Tertiary and Quaternary sediments.

Glacial moraines extend out several miles beyond the edge of the mountains, surrounding flat-topped mesas of Tertiary (Oligocene) sand and gravel. On this side of the Uintas, valley glaciers fanned out at the base of the mountains.

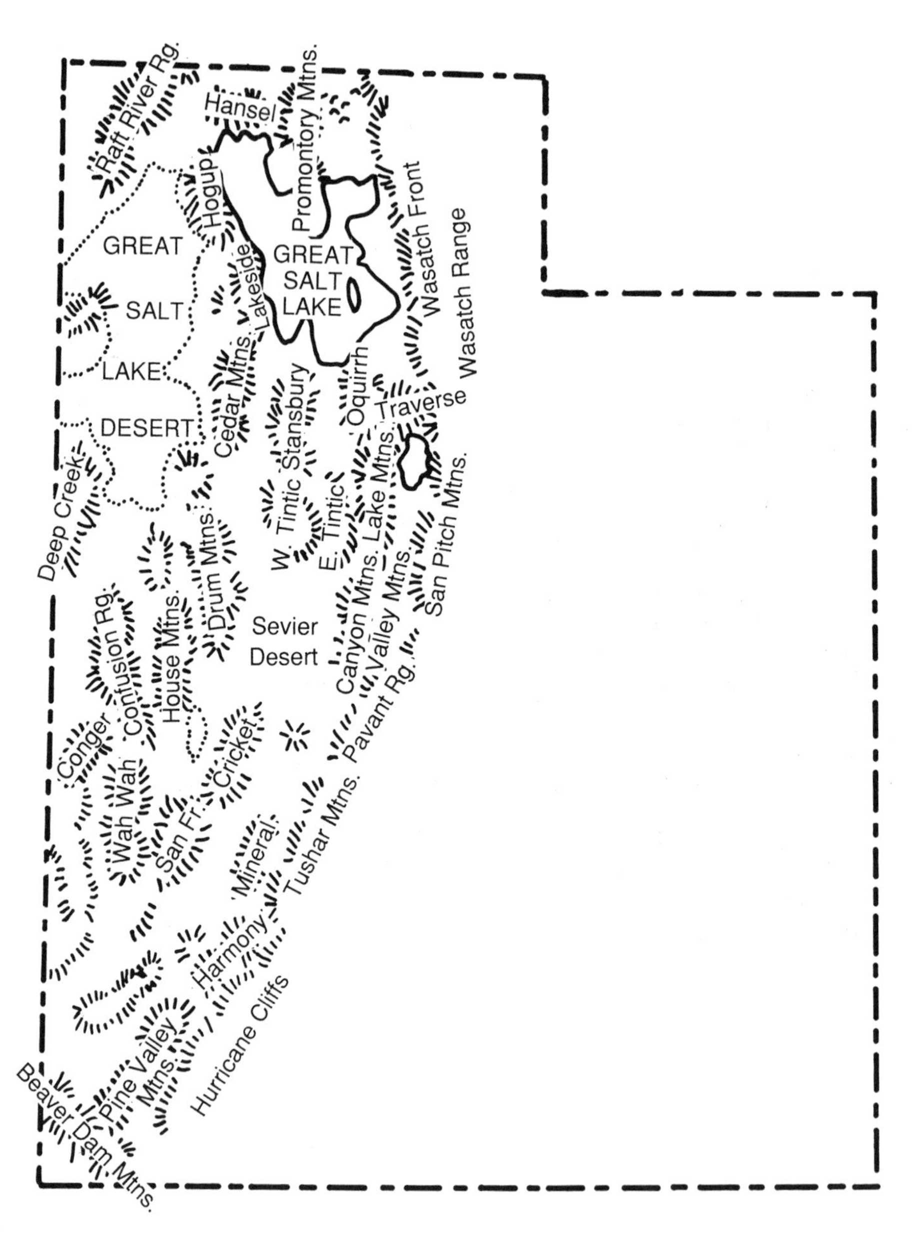

Utah's share of the Great Basin contains many individual mountain ranges, most of them trending north-south. Not all shown at this scale, the ranges are separated by arid desert basins, some with drainage to the outside, others with flat, salty playas marking their lowest points.

III
The Great Basin and the Wasatch Front

Utah's western desert is geologically and geographically part of the immense region known as the Great Basin. Here, in a land of salt and sun and gaunt mountain ridges, no streams, no rivers ever reach the sea. Except for that which evaporates in the dry air, rain and snow that fall in the Great Basin sink into the gravel and sand of valley floors. Only in the extreme southwest corner of the state do a few intermountain valleys drain south toward the Colorado River, thereby ultimately reaching the Gulf of California.

The Great Basin in turn is part of the Basin and Range Province, a region characterized by rank upon rank of long, narrow mountain ranges, tilted fault blocks of the Earth's crust, that alternate with intermountain basins partly filled with gravel and sand chiseled from the mountains. In Utah most of the ranges are oriented about north-south.

Many intermountain basins also contain deposits of salt accumulated in lakes and pools, then concentrated by the desert sun. Brilliant white playas today mark the lowest parts of non-draining basins. Storm-fed streams only occasionally reach them, bringing dissolved solids — calcium carbonate, common salt, and other minerals picked up from rocks and soil. As the playa ponds evaporate, these minerals remain behind, each time adding a shining new layer to the white playa surface.

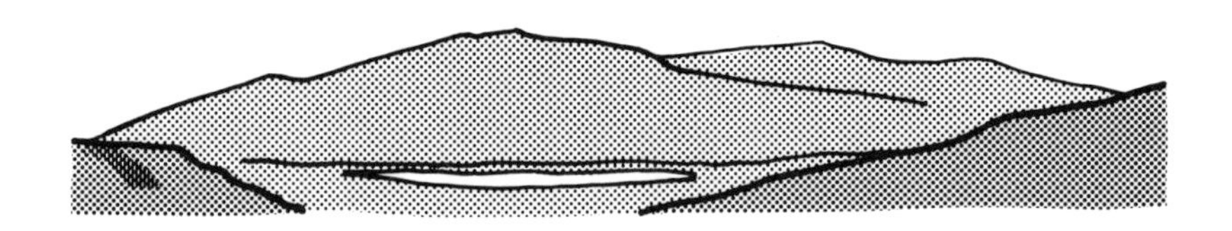

The shimmering playas of undrained intermountain basins are among the flattest surfaces on Earth.

Many playas exist in western Utah. During the 1980s, a period of unusually heavy precipitation, several of them once more became real, albeit shallow, lakes. Utah's largest playa, 40 miles across from east to west and 100 miles from north to south, lies between Great Salt Lake and the state's western border: the Salt Lake Desert or Bonneville Salt Flats. Part of its smooth, flat surface — one of the flattest places on Earth — has attained fame as the Bonneville Speedway.

Great Salt Lake itself, described more fully in the next section, lies in the largest and deepest of the intermountain basins, close to the Wasatch Range. It was long thought to be gradually drying up, but the wet cycle of 1983-87 showed emphatically that it is not.

Most of the time, in both draining and undrained intermountain basins, streams from surrounding mountains sink into porous desert soils of the basins. Then, as this soil water evaporates, it leaves its dissolved solids behind, gradually forming caliche (ca-LEE-chee), a white or dirty gray mix of calcium carbonate and other minerals deposited around and among sand grains and pebbles. As it cements the sand and pebbles, caliche reduces the permeability of the soil, retarding both erosion and the growth of plant roots.

A number of geologic and climatic events, overlapping in time and space, is responsible for the gaunt ranges and arid, sun-baked basins of Utah's western deserts:

• Development of the Sevier Mountain belt in Cretaceous time, with movement on large eastward-directed thrust faults narrowing Utah by 40 miles or more;

- Igneous activity — abundant intrusions and huge volcanic outbursts — in mid-Tertiary (Eocene to early Miocene) time;
- Mid-Tertiary regional uplift, with Utah and adjacent states domed upward as much as 5000 feet;
- Mid-Tertiary to present-day crustal stretching, resulting in normal and detachment faulting and creating the linear mountain ranges and desert basins;
- Simultaneous infilling of intermountain basins with rock and sand chiseled from the mountains, and with volcanic outpourings;
- Development of Lake Bonneville in Pleistocene time, its waters reaching all the way to the western and northern margins of the state;
- Increasing aridity in the last 10,000 years, with consequent drying up of most streams and lakes and marked reduction in the size of Lake Bonneville. As its level dropped below its outlet, the lake became salty.

Many features of the Great Basin region are unique to deserts. Rain is sparse, but weathering and breakdown of rock continues in the mountains even without it: Overnight frost breaks the rocks apart, gravity tugs at steep cliffs, plant roots pry rocks apart, wind hammers at the rocks with grains of sand. Such breakdown provides more than enough rock material to overload the occasional torrential downpours and the born-again streams that they engender. Heavy rains instantly convert the usually dry desert washes into roiling torrents, the fabled "walls of water" of western fact and fiction. These sudden flash floods carry heavy loads of rock debris.

Overburdened streams deposit their loads, rounded into pebbles and boulders or broken into sand grains, where their courses become less steep: at the mouths of their canyons, along the mountain fronts. There, pebbles and boulders and sand accumulate in alluvial fans, commonly merging in alluvial aprons or bajadas (ba-HA-das) that extend toward the centers of the valleys, there meeting with counterparts from neighboring ranges.

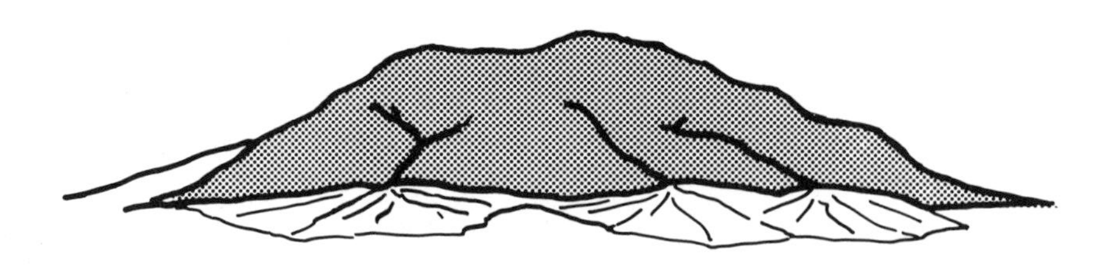

Alluvial fans at the mouths of mountain canyons may merge to form alluvial aprons or bajadas around the desert ranges.

Wind erodes the desert, too. Desert dust storms, as well as the many dust devils that spiral against the summer sky, remove tons of sand and silt and clay from desert surfaces. Removal of fine materials often leaves behind a layer of pebbles known as desert pavement, which then armors the desert against further erosion by wind or water.

Most of the fault-block mountains of the western deserts seem to have begun their development in Oligocene and Miocene time, 35-24 million years ago. Once started, faulting proceeded at a rapid rate — not just a neat pattern of slices, but with new faults cutting across old ones, and with the belt of new faulting moving gradually east. Fault movement continues:

As mountain fronts erode in arid regions, they slowly recede, leaving lightly graveled pediments of solid rock that merge smoothly with surrounding alluvial fans and bajadas.

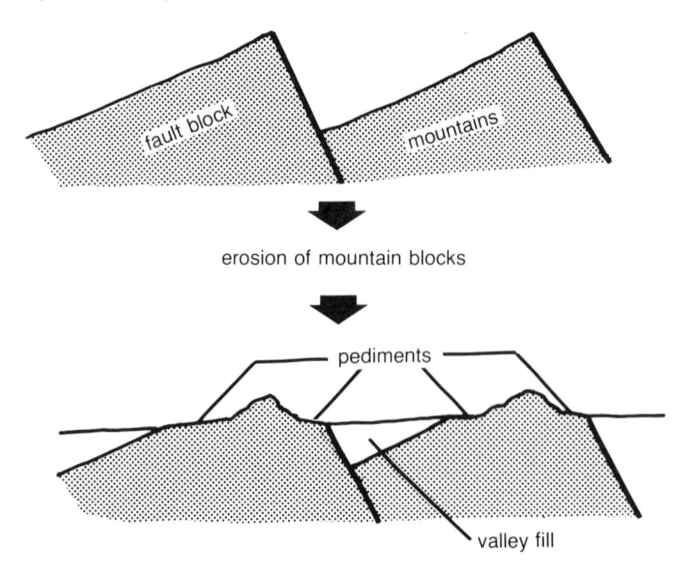

As wind removes dust and sand, pebbles remain behind, forming desert pavement.

Some of the ranges are still growing at an average rate of about 2.5 inches every 100 years.

One of these faults, the Wasatch fault, is so pronounced that even in everyday English its high, prominent scarp is called the Wasatch Front. Earthquakes with epicenters along this great fault show that it is still active. A great part of Utah's population lives in the shadow of this fault, whose sudden movement could cause a truly disastrous earthquake.

The mountain-sized fault blocks of this region developed in a fairly orderly fashion, breaking roughly at right angles to the westward direction of stress. The stress itself must be due to deep-down convective movements in the mantle, which have dragged the southwestern part of the continent over an old East Pacific plate and are now rubbing it against the Pacific plate along the San Andreas fault of California.

Most of the breaks between blocks are north-south-trending normal faults that curve and flatten out with depth. Many seem to merge several miles down with others to become undulating detachment faults. The separate blocks, north-south slivers of crust, rotate against these curving faults, so their west edges are high, their east edges low. Because in most cases their low east sides are now buried beneath valley fill, the mountain blocks appear to be faulted on their west sides only.

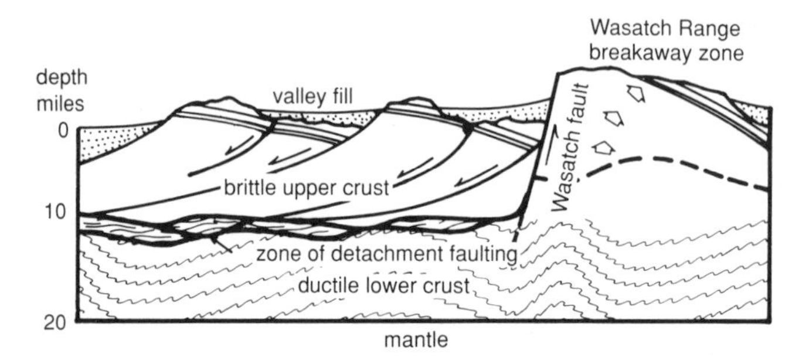

High-angle normal faults are now thought to curve at depth, some of them perhaps merging with detachment faults that underlie the Great Basin. The Wasatch Mountains are the breakaway zone for the Wasatch fault: sedimentary rocks that once covered this area have slid westward. Relieved of their burden, the Wasatch Mountains have pushed upward strongly — and are still doing so.

As geologists work out the patterns of individual ranges and interpret seismic data from intervening basins, it becomes increasingly apparent that the continent is being pulled apart here. Seismic records suggest that detachment surfaces lie about ten miles down, and extend under almost all of the Great Basin. Detachment faulting has about doubled the width of the Great Basin. The Earth's crust is much thinner and hotter here than in other parts of Utah or for that matter in other parts of North America. Its lower portion, 5-15 miles down, resting on the mantle, behaves as if it were plastic or ductile; it can be stretched or drawn out horizontally like taffy, given enough time. With the deep crust stretching out, the upper brittle crust breaks into slivers oriented at right angles to the stresses. As we've seen, these slivers are now the desert mountain ranges.

In just a few of Utah's western ranges, as in others of Nevada, California, and Arizona, detachment faults now appear at the surface as the rounded summits of metamorphic core complex mountains. The rise of these "turtleback" mountains is credited, in part at least, to removal of the thick blanket of sedimentary rocks that slid off during detachment faulting.

With their overburden removed, the unroofed cores of these mountains pushed upward, often in hingelike fashion, revealing the detachment fault surfaces as carapaces of broken, crushed, partly melted metamorphic rock smeared out by the friction and heat generated by fault movement. Movement may have been aided and abetted by heat associated with Miocene volcanism and the rise of many igneous intrusions.

A few of western Utah's ranges contain radioactive minerals. Many contain ores of gold, silver, iron, copper, and other valuable substances. Almost without exception these minerals occur where Tertiary igneous intrusions penetrated Paleozoic or Mesozoic limestone, bringing in, as fluids, concentrated mineral solutions. With the limestone acting as a catalyst, the ore minerals precipitated from these solutions, enriching both the limestone and the igneous rocks of the intrusions.

Mineral-hunting is good in most of the mining areas, especially on dumps near the mines. But be very careful. Old mines are extremely dangerous. Stay out of them. Rockhounds should use considerable care in choosing collecting sites, and of course should obtain permission before searching on private property.

Roads are few in the vast reaches of Utah's western desert. Only five highways cross it from east to west, all but Interstate 80 sparsely traveled. One north-south highway, Interstate 15, pretty well defines its eastern edge; there are no other north-south routes. In this chapter, the geology of the Wasatch Front and the Hurricane Cliffs, the edge of the mountains, is included in the logs for Interstate 15.

Great Salt Lake

Certainly Utah's largest single geologic attraction, Great Salt Lake also plays a role in several of the state's major industries. Prehistoric man lived by its shores and utilized its products, particularly salt and the abundant supply of edible-rooted cattails in marshes near river mouths. And modern man uses its salt, as well as its plentiful brine shrimp as food for his tropical fish. The lake also serves, of course, for recreation — both swimming and boating.

Because it has no outlet, the lake is salty: It contains salt, gypsum, magnesium, potash, and many other dissolved substances carried into it by streams and rivers and concentrated by evaporation of the lake waters. The saltiness varies quite a bit with changes in lake level, but in historic time it has consistently been far saltier than sea water. Its level and its saltiness depend upon rates of stream flow, rainfall, and evaporation. In 1963, the year of the lowest historic level, dissolved salts reached a maximum of 28.5% by weight, nine times saltier than sea water. In 1988 (as this book is written), after four years of above normal precipitation, the lake is at its historic high and the salt content is down to about 13%.

The lake is divided into two parts by a railway causeway between Promontory Point and Lakeside. Even though there are openings that allow exchange of water, the part of the lake south of the causeway, which receives almost all of the freshwater inflow, is less salty than the part north of the causeway. The lake is fairly shallow as lakes go, varying from 27 to 47 feet maximum depth between historic low and high water levels.

The lake was not always salty. Its ancestor, a Pleistocene lake which geologists call Lake Bonneville, had an outlet to the north into the Snake River drainage of Idaho. Abundantly fed

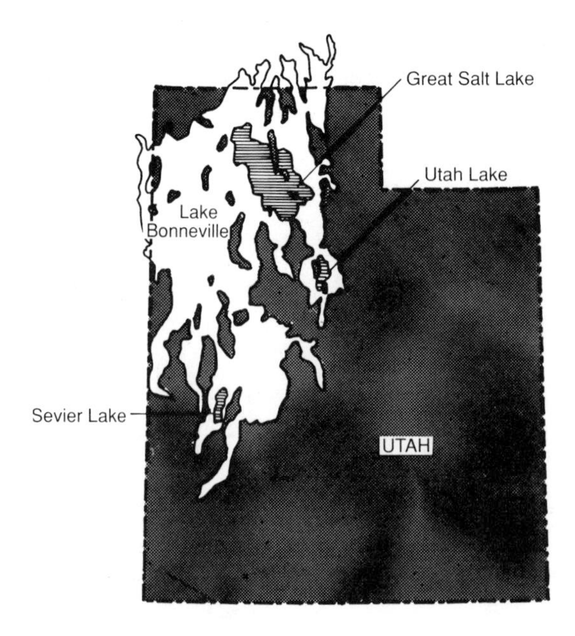

Fed by the abundant precipitation of ice-age time, Lake Bonneville extended far beyond the limits of today's Great Salt Lake. Many of the present mountain ranges were islands in its broad expanse.

by mountain glaciers and the heavy rainfall and snowfall of ice-age time, Lake Bonneville was then about 1100 feet deep; it spread out over much of western Utah and even reached into Nevada and Idaho.

The present lake occupies three parallel, interconnecting grabens, down-faulted strips of the Earth's crust, delineated by irregular rows of uplifts that form islands and promontories, mountain ranges with their feet in the water. Both the grabens and the uplifts that separate them trend about 20° west of north, parallel to the front of the Wasatch Range.

Geologists believe that under the lake's waters and bottom sediments, the depressions in which Great Salt Lake now lies are no deeper than other intermountain basins in this region. The lake is here instead of somewhere else because streams that drain the resistant crags of the Wasatch Range just don't carry as much sand and gravel and rock as do streams that drain mountains to the north and south, composed of more easily eroded rocks.

The earliest lakes to occupy these three low grabens probably developed in Miocene and Pliocene time — well before Lake Bonneville. They are thought to have been freshwater lakes. We don't know much about them because the only evidence that they existed is now hidden in the bottom sediments of Great Salt Lake, which have not been extensively sampled and studied. A major change came, though, between 130,000 and 30,000 years ago, when volcanic eruptions in southeast Idaho blocked the Bear River's earlier course to the Snake River, forcing it to turn southward into the Great Salt Lake depression.

In Pleistocene time, snowfall increased and glaciers developed in Utah's mountains. In the Great Basin, rainfall also increased. And in the cooler climate evaporation was reduced. Flooded streams and turbulent rivers carried more and more water into the basins west of the Wasatch Range. By about 25,000 years ago, whatever lakes were here had merged, deepened, and overflowed into adjacent basins, creating Lake Bonneville. Today's Great Salt Lake Desert and Sevier Desert were inundated as arms of the lake spread westward and southward. The present ranges of western Utah became islands in the vast expanse of water; present islands were submerged. Eventually, around 16,000 years ago, the rising lake overflowed northward through a narrow outlet near Downey, Idaho.

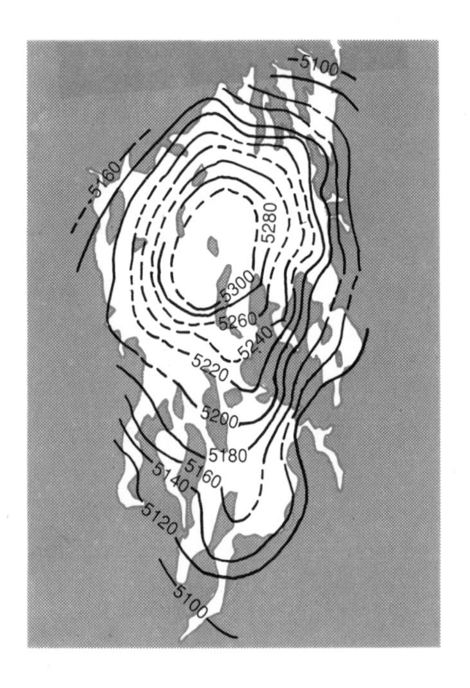

By mapping the present elevations of Lake Bonneville's shoreline, geologists show that the weight of its water substantially bowed down the Earth's crust. Present rebound is as great as 200 feet.
—Adapted from Crittenden, 1963

Lake Bonneville remained at or near the level of its northern outlet — about 870 feet above the historic (since 1850) level of Great Salt Lake — for about 1500 years. During that time, its waves carved horizontal benches into the surrounding mountains — the highest old shorelines that can be seen on the west face of the Wasatch Range, as well as on the various islands and promontories of the present lake, and on the flanks of desert ranges far north and south and west of the present lake. These uppermost shorelines are now collectively known as the Bonneville shoreline.

Then, about 14,500 years ago, the lake breached its threshold. Lake waters raced and tumbled down the former exit route as a more-than-catastrophic flood, churning northward into the Snake River in Idaho. You can imagine the magnitude of the thundering waters: In less than a year, over the whole vast extent of Lake Bonneville — some 19,800 square miles — lake level was lowered by 350 feet! A new threshold was finally established in resistant bedrock at what is now Red Rock Pass, just across the state border in southern Idaho.

The lake stayed at the level of this new threshhold for about 1000 years — from 14,500 to 13,500 years ago. During this time the Provo shoreline developed. Broader and more obvious than the Bonneville shoreline, the Provo shoreline includes, in many places, wide benches cut into the mountains and islands, covered with gravel and fringed with deposits of tufa, calcium carbonate deposited by the lake waters. It also includes beach deposits, spits, and, near the mountains, large deltas built out into the lake. The level platforms and flat-topped deltas make handy and scenic building sites, and parts of several of Utah's cities perch on the Provo benches.

As the Provo shoreline developed, the threshold at Red Rock Pass was gradually eroded; a later set of beach ridges occurs about 12 feet below the main Provo level.

Eventually, the lake dropped well below the Provo level, seemingly because of the increasingly dry climate as the ice ages drew to a close. Periodic stable levels were established by a balance between inflow into the lake and evaporation and groundwater flow out of it. At times the lake dropped to the present level. Between 11,000 and 10,000 years ago, the lake became salty.

The lowest shorelines, developed in just the last 10,000 years, are fairly obscure low-relief benches, ridges, and gravel bars. Around 7500 and 5000 years ago the lake was apparently even lower than it has been in historic time. Its floor today is remarkably smooth and flat, a feature that suggests that it dried up completely for a time, leaving a salt-floored playa similar to the dry lakebeds of other Great Basin valleys today. Aerial photographs show huge underwater mud cracks like those that develop in some playas today. The cracks outline areas up to 330 feet across, and are much larger equivalents of the small shrinkage cracks we see today on drying mud puddles.

In historic time the lake's level has ranged from 4191 feet to 4217 feet above sea level — a range of 26 feet. Normally, it fluctuates two to three feet with seasonal changes in rainfall

A canal carries water pumped from Great Salt Lake into a desert basin to the west. After evaporating half of its volume, the lake brine is allowed to flow back into Great Salt Lake, preventing buildup of salt in the adjacent basin.

and snowmelt. Between 1983 and 1987, however, it rose 12 feet, flooding roads, railroads, industries, and recreation areas along its shore. As the lake reached its historic high in 1987, a new pumping station near Lakeside, on the western shore, began to pump lake water into a dry desert basin west of the lake. The new lake created there increases the evaporating surface and, it is hoped, will prevent higher lake levels that would cause further or more severe damage to the works of man.

Rivers entering the lake form broad mudflats and birdfoot deltas on which they branch and rebranch in crooked channels. Salt flats have developed in many places, particularly on the western side of the lake where no rivers carry in sediment. Oolitic sand, with rounded, light-colored grains of calcium and magnesium carbonates, borders much of the lake. The oolites — the individual grains — form in agitated water as these carbonates build in layers around nuclei of some kind — commonly, in Great Salt Lake, around fecal pellets of brine shrimp. Along the east shore particularly, the natural shoreline has been much altered by man.

The same rivers as they enter Great Salt Lake bring in the dissolved solids that make it salty. As long as the lake has no outlet, its saltiness will gradually increase. In addition to regular table-variety salt (sodium chloride), the lake contains silica, potassium, nitrate, sodium bicarbonate, magnesium, calcium, sulfate, and other dissolved substances. All these solids are concentrated by evaporation of lake waters. Some of them are probably hand-me-downs from Lake Bonneville, concentrated ever since that lake dropped below its outlet threshold at Red Rock Pass in Idaho.

Early Indians, as well as trappers and immigrants, obtained salt from the lake, as did Mormon pioneers. Salt production is still a major industry. By controlling the flow of brine in and out of settling ponds and evaporating pans, salt producers obtain various substances. Water from the lake is first pumped into settling ponds along the lake shore to allow insoluble materials to settle out. Then moved to evaporating pans, it is concentrated by natural solar evaporation and dry desert air until calcium and magnesium carbonates crystallize and settle out.

Along its unfrequented north shore, several large freshwater springs, recognizable by dense vegetation close to the water, feed into Great Salt Lake. Few plants can grow along saltier parts of the shore.

At this stage, the "salt point," the brine is pumped into other ponds, where with further evaporation common salt precipitates. Remaining brine is pumped into yet other ponds and allowed to evaporate further to precipitate magnesium and potassium.

Once harvested by labor-intensive plowing and hand-shoveling, the salt is now collected by mechanized harvesters that scrape up the top 4-6 inches from the evaporating ponds. After harvesting, some of the salt is processed to remove impurities; the rest goes as it is to salt highways in snowy weather or to chill brines used in making ice cream.

What of the future? Studies of climate cycles suggest that the level will continue to fluctuate between 4200 and 4218 feet above sea level for another 150 years. Then, say long-range forecasters, it will rise above its 1987 historic high of 4217 feet.

Interstate 15
Idaho—Salt Lake City
91 miles/147 km.

Interstate 15 follows the Malad River southward from the Idaho border, across sediments deposited in Lake Bonneville. Hills to the west consist of Pennsylvanian/Permian limestone and quartzite that are part of the Sevier thrust belt, displaced eastward 40 miles or more. Farther west are the summits of the Raft River Range, one of Utah's metamorphic core complexes. These mountains consist of Precambrian sedimentary rocks that once were below the Basin and Range detachment faults. Lightened by removal of thick layers of their sedimentary rock blanket, which slid eastward along the detachment faults, the range rose in Miocene time.

Mountains east of the highway rose along the Wasatch fault, which pretty much parallels I-15 all the way to Scipio, a distance of about 200 miles. This fault, actually a zone of branching and reuniting faults, is the easternmost large Basin and Range fault; uplift along it measures many thousands of feet. Movement on the Wasatch fault began in Miocene time; frequent earthquakes with epicenters along the fault show that movement continues today.

Lake Bonneville carved its horizontal shorelines on the mountains and hills of western Utah.

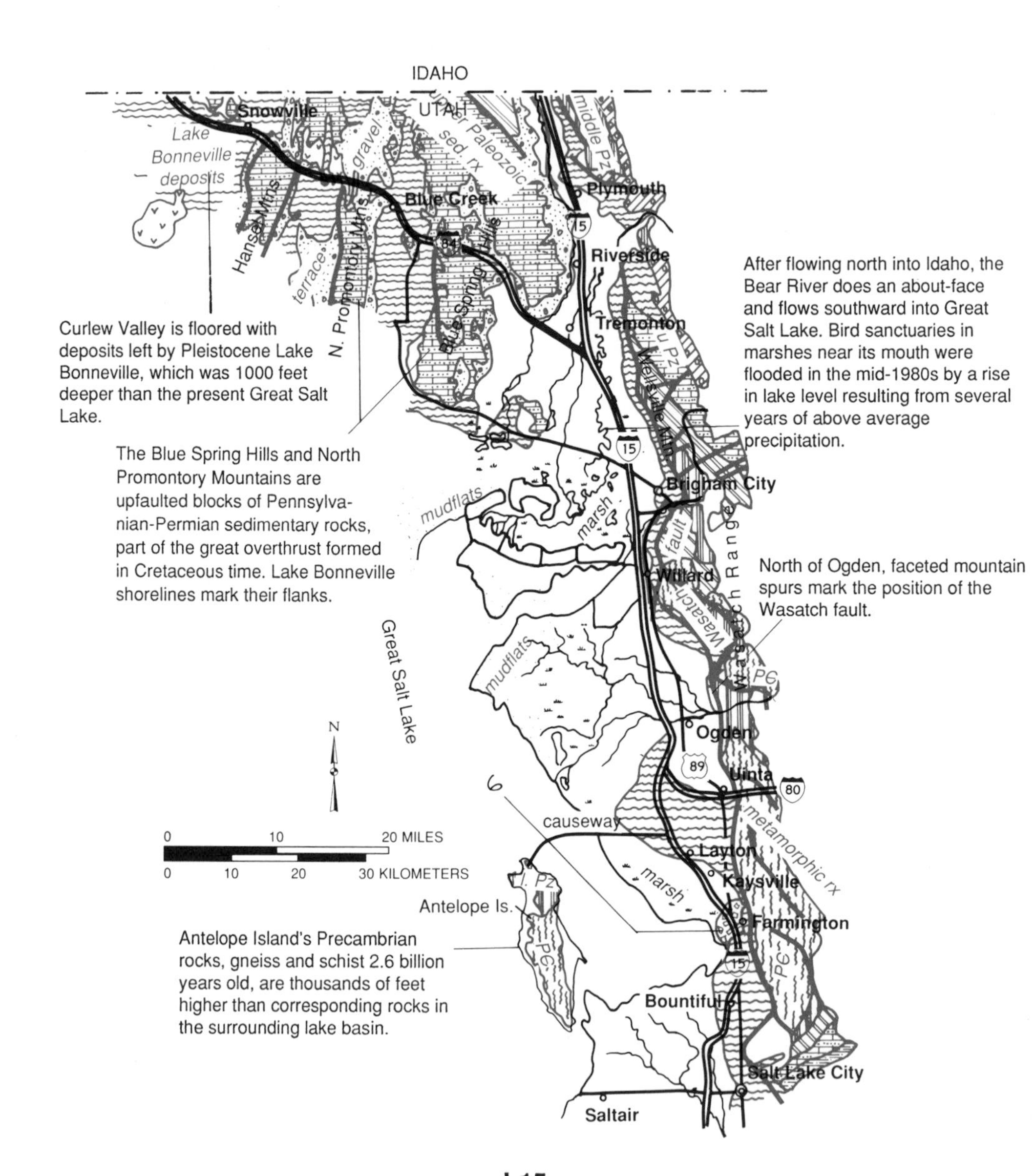

I-15
IDAHO — SALT LAKE CITY

I-84
IDAHO — I-15

The Bear River, after looping northward from Bear Lake, enters the Malad River valley southeast of Plymouth. It does not join the Malad River, however. It turns south on the extremely flat valley floor, a heritage of flat-lying sediments deposited in Lake Bonneville, and parallels the Malad for some distance, winding back and forth in intricate meanders. The Bear and Malad rivers finally lose themselves in the marshes of Bear River Bay, an arm of Great Salt Lake.

Lake Bonneville shorelines crease mountain slopes all along this route. Neither the uppermost or Bonneville shoreline nor those immediately below it are particularly deeply etched. But some distance below them is the prominent Provo shoreline, in places expressed as broad terraces that were once deltas of streams entering Lake Bonneville. Parts of the interstate highway are on the terraces. Near Honeyville, an alluvial fan formed in Bonneville time reaches out into the valley. Similar fans, as well as other flat-topped deltas whose surfaces correspond to the Provo level, can be seen farther south. In several of the towns and cities along the mountain front, high delta surfaces on the Provo shoreline have become choice residential areas.

Wellsville Mountain east of Honeyville is a single large upward-faulted ridge of Precambrian metamorphic rocks covered with layered Paleozoic sedimentary rocks. The Precambrian rocks, about 1.2 billion years old, appear at the bottom of the slope; they include large amounts of phyllite, a type of schist especially rich in mica.

Higher on the slope is the resistant, cliff-forming Brigham quartzite. This hard, tightly cemented Cambrian rock, deposited originally as sand on beaches and bars as a shallow western sea crept across the continent, correlates with Cambrian sandstones in states south, north and east of Utah. Because encroachment of the Cambrian sea was gradual, advancing from west to east, the sandstones become younger eastward. Everywhere they mark the beginning of Paleozoic sedimentation, the beaches and bars at the margin of the advancing sea.

Above the Cambrian sandstone are two other Cambrian units: a thick layer of shale and shaly limestone, and above it limestone and dolomite — a three-part sequence typical for Cambrian rocks in all of western North America. The sequence reflects the increasing depth of the Cambrian sea and the increasing distance from shore, the source of sand, as the shoreline moved eastward. Younger Paleozoic rocks, Ordovician to Permian in age, appear on the crest and east side of the range.

A section across Wellsville Mountain appears with the roadlog for US 89 Brigham City to Garden City on page 154.

A prominent break in the slope of Willard Peak, just north of Ogden, marks the position of the Willard thrust fault, a major fault of the Sevier thrust belt, formed in Cretaceous time. The fault plane tilts east, and earlier geologists concluded that the upper plate, above the fault, had moved west — opposite the known movement on other Sevier thrust faults. However, study of associated thrust faults, as well as of drag folding of rocks below it, shows that movement was east, in keeping with other Sevier faulting, but that the fault plane was tilted during later movement on Basin and Range faults.

Near Ogden, both railroad and highway run for some distance barely above lake level. During the high lake levels of the mid-1980s the lake's waters lapped against the railroad and highway berms. The marshy region west of the highway is part of the delta of the Bear River, where its waters, and the sand and silt they carry, have partly filled in Bear River Bay. The marshes include the Bear River National Wildlife Refuge, also engulfed by the 1980s rise in lake level.

The mountains developed during the Basin and Range faulting of Miocene and later time. Mountain spurs near and south of Ogden are truncated, or faceted, cut off crosswise so that they end in triangular faces or facets. Such facets indicate quite recent uplift on the Wasatch fault. They also show the exact position of the fault.

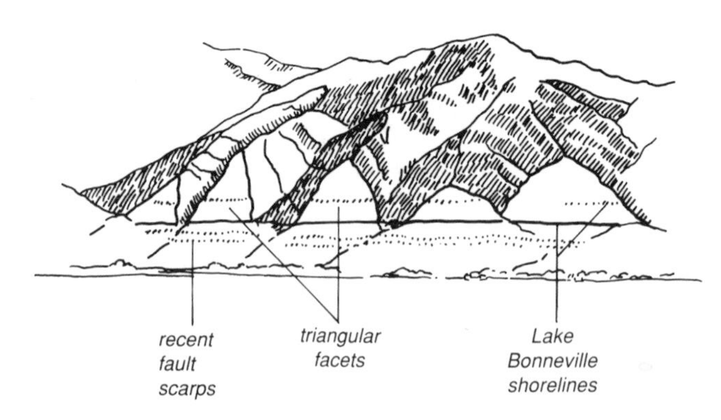

Triangular facets at the lower ends of mountain ridges indicate continuing movement on the Wasatch fault. Erosion has not yet sharpened the ridges.

Several miles south of Ogden, the highway rises onto another Lake Bonneville delta, this one formed by the Weber River. Layton and Kaysville are on this delta, and US 89 crosses it between the interstate and the mountains. With headwaters in the Uinta Mountains, the Weber River was particularly well fed in Pleistocene time, when the high western part of the Uintas was covered by one big icecap. The river is one of the few that flow completely through the Wasatch Range.

Quarries here and farther south produce gravel from the Weber River delta. At the base of the mountains, quarry excavations have exposed the grooved and polished surface of the Wasatch fault.

Near Farmington, debris from a large landslide several miles across extends out into Great Salt Lake. South of Farmington the delta terrace narrows; no large stream canyons cut the massive Precambrian rock that makes up this part of the Wasatch Range, and without streams no deltas developed along the shores of Lake Bonneville. The easternmost tip of Great Salt Lake reaches almost to the mountains here where the hard Precambrian rocks contribute little to the basin fill that elsewhere pushes the lake shore west.

South of Bountiful, south of the erosion-resistant Precambrian part of the Wasatch Range, streams again become plentiful. They flow through Paleozoic sedimentary rocks that erode more easily than the Precambrian massif. These streams have built an apron of merging deltas that extends south of Salt Lake City to merge with the delta of the Jordan River, which in turn flows from Utah Lake into Great Salt Lake. Like towns farther north, Salt Lake City initially developed on the delta deposits; it has now spread west across what was once the floor of Lake Bonneville.

The Wasatch fault swings west and crosses these deposits west of the main part of the city. Hot, salty springs occur along this part of the fault — another indication that this region is still geologically active.

Salt Lake City, like other cities just west of the mountains, suffers from a perennial water shortage. Streams draining the west face of the Wasatch Range, as well as those whose natural drainages bring water through the range from higher terrain to the east, do not furnish enough for the needs of the densely populated area west of the mountains. A complicated network of dams, canals, pipelines, and tunnels now augments the flow of several rivers, the Spanish Fork and Provo rivers in particular, by diverting Uinta Mountain streams that formerly drained into the Green River and ultimately into the Colorado River and the Gulf of California.

The highly populated area just west of the mountains is also a highly seismic area, with strong earthquake risks. It's sort of scary when you realize that medium-sized quakes recorded along the Wasatch fault zone occur at an average rate of about one a year, and that much stronger quakes must have accompanied the thousands of feet of movement on the Wasatch fault. Epicenters of other quakes string out along corresponding faults farther south. Loosely

consolidated delta gravels and clayey Lake Bonneville deposits are likely to slump during severe earthquakes, presenting a danger to the whole urban corridor. Many buildings in Salt Lake City are known to straddle branches of the Wasatch fault. The mountain front is also dangerously steep here, rising threateningly above the towns and cities, as if waiting for a severe earthquake to shake loose a really large landslide similar to the one near Farmington. Significant downward movement west of the Wasatch fault could bring Great Salt Lake waters surging into low-lying parts of the city as well.

Oil refineries on the outskirts of Great Salt Lake process oil from eastern Utah. The mountains beyond harbor the world's largest open pit copper mine near Bingham.

Interstate 15
Salt Lake City—Spanish Fork
55 miles/88 km.

Backed by the steep western face of the Wasatch Range, and with the blue waters of Great Salt Lake stretching far to the west, Salt Lake City is one of our most scenically sited cities. Its scenery is its geology. But here geology presents problems. The rising and falling of lake levels threaten lakeside industries and recreational facilities. The Wasatch fault, right at the base of the mountains, brings the threat of earthquakes. The steepness of the mountain front presents the possibility, even probability, of massive, earthquake-triggered landslides.

In the southern part of the city, a branch of the Wasatch fault, the East Bench fault, runs through several residential districts. With a more or less north-south trend roughly paralleling 13th East Street and Highland Drive, the fault cuts through and offsets layers of porous, water-bearing gravels, Lake Bonneville deposits. Normally, water from the mountains, sinking into the sloping terrace gravels, flows as groundwater toward Great Salt Lake, held down by sloping, impermeable layers of shale. As more water is added near the mountains, that in the gravel layers develops considerable hydrostatic pressure. Where the seal of the shale layers is broken by faults, water reaches the surface in artesian springs. A line of such springs led to discovery of this branch of the Wasatch fault. Boggy ground along the line of springs has in the past led to failing building foundations. In the area near Big Cottonwood and Little Cottonwood creeks, east of mileposts 301 and 298, the Wasatch fault cuts and displaces, by as much as 50 feet, some of the Lake Bonneville shoreline features, revealing that the fault has been active since these shorelines developed.

Mountains visible to the west from interchange 302 are the Oquirrh Mountains, composed largely of Pennsylvanian-Permian limestone and sandstone of the Oquirrh group. Some 25,000 feet of these sedimentary rocks exist in the Oquirrh Mountains, five vertical miles of limestone and quartzite! The big copper pit at their southern end, where Tertiary granite intrudes Paleozoic rocks, is 2.5 miles in diameter — one of the world's largest open pit mines. Only the mine dumps are visible from the highway. Copper ore, discovered in 1906, occurs

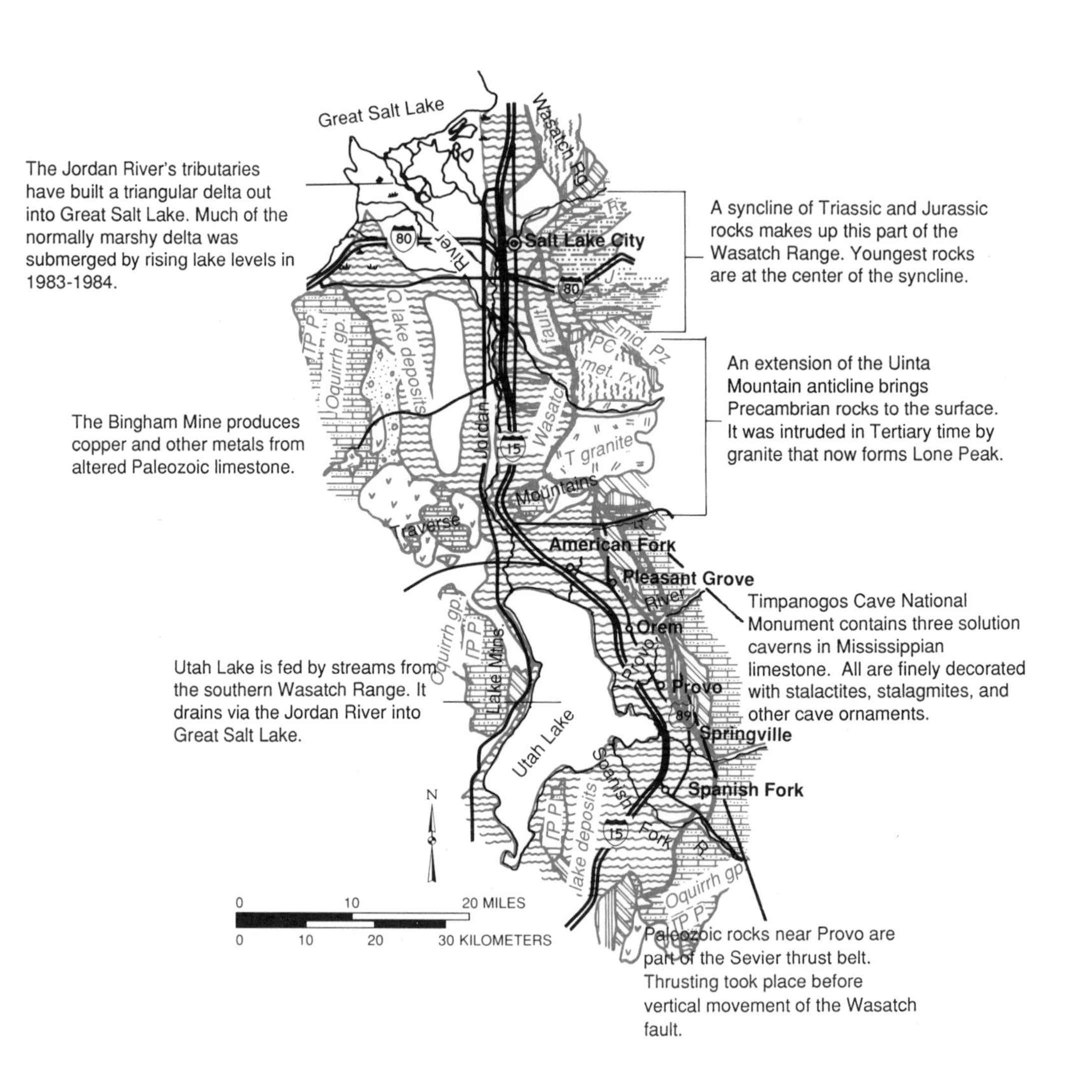

I-15
SALT LAKE CITY — SPANISH FORK

where the intrusion and metallic fluids associated with it altered and enriched Paleozoic limestones, suffusing them with metallic minerals. In addition to copper, the ore contains molybdenum, gold, lead, and zinc. The mine has been America's most productive copper source. Half a million tons of rock, of which about a fifth is ore, are excavated every day.

Large cement and gravel works near Draper, milepost 291, offer a good look at the delta gravels along the Wasatch Range, their pebbles and cobbles rounded and well sorted by ice-fed streams and the lapping waters of Lake Bonneville.

The Traverse Mountains, an east-west range clearly marked with Lake Bonneville shorelines, closes the southern end of the Jordan River's valley. These mountains are geologically continuous with the Oquirrh Mountains block — a sort of crossbar connecting it with the Wasatch Range. The Oquirrh group sedimentary rocks extend across the Jordan River and under I-15.

Utah Lake is a freshwater lake fed by the Provo River and streams draining the Wasatch Range. It drains into Great Salt Lake via the Jordan River, adding its small quota of minerals to the larger lake. Its waters are used for irrigation in Salt Lake Valley to the north.

The lake occupies a graben dropped along faults between the Wasatch Range and the Lake Mountains. In Pleistocene time the graben held an arm of Lake Bonneville; shorelines can be seen on both the Traverse Mountains and the Wasatch Range. Some 13,000 feet of lake sediments lie beneath the central part of Utah Lake's valley.

Also visible from the interstate are the truncated spurs of the Wasatch Mountains, evidence that movement on the Wasatch fault is geologically recent. Studies of the fault in relation to the Lake Bonneville shoreline show that it has been active since the time of the lake's maximum extent.

American Fork, Orem, Provo, and Spanish Fork are all on Lake Bonneville deposits. Near Orem and Spanish Fork these consist of delta gravels from the Provo and Spanish Fork rivers. The ancient shorelines show particularly well near Orem.

Iron smelters and steel mills near Provo process ores from Wyoming. Ranges east of Provo and Spanish Fork are capped with Paleozoic rocks, including the Oquirrh formation, that moved eastward over Jurassic and Cretaceous rocks as part of the Sevier thrust belt. These mountains, too, show the Lake Bonneville shorelines.

The Pleistocene delta of the Spanish Fork River, near the town of the same name, is particularly large; it spreads well out into the

valley, partly restricting the size of Utah Lake. The river was fed by glaciers in Pleistocene time, which accounts for the size of the delta. It is formed by the union of Thistle Creek and Soldier Creek, both of which flow through relatively soft, slide-prone Tertiary sediments, another factor favoring rapid delta development. Gravel pits near Spanish Fork show that pebbles and sand of the delta gravels are poorly sorted, with little stratification, a sign of very rapid deposition.

Interstate 15
Spanish Fork—Scipio
62 miles/100 km.

The steep, abrupt front of the mountains east of Spanish Fork is the scarp of the Wasatch fault. Mountain spurs, which near the summits are sharp-edged ridges, are cut off by the fault, leaving large triangular facets along the lower mountain slopes. These facets are evidence of fairly recent movement; a long period of erosion would have resharpened the ridges.

Mountains east and south of Spanish Fork consist almost entirely of Pennsylvanian and Permian limestone and shale of the Oquirrh formation. Deposited in a deep marine basin farther west, this rock moved eight to ten miles eastward as part of the Sevier thrust belt. Similar strata form little West Mountain, west of the highway. South of Payson the lower mountain slopes contain older Paleozoic strata, Mississippian down to Cambrian.

Wave-cut shorelines of Pleistocene Lake Bonneville show up well on these mountains. Each shoreline developed during a time when the lake level remained the same, controlled by hard rocks just north of the Idaho line, where the lake overflowed into the Snake River drainage. When the lake was at the Provo level, the Spanish Fork River built a sizeable sand and gravel delta out into it, a delta readily identified between Interstate 15 and the mountains. Finer lake sediments deposited after Great Salt Lake became salty form the broad terrain that slopes toward Utah Lake.

South of Spanish Fork, the Wasatch fault and its line of truncated spurs swing sharply westward, forcing the highway westward also. Pleistocene mountain glaciers carved high cirques and gouged the valleys on Loafer Mountain southeast of Santaquin as well as on Mt. Nebo, still farther south.

The highway rises to the Provo shoreline near milepost 256. Well to the west are the Tintic Mountains, site of the Tintic lead- and silver-mining district. The ores there are associated with a Tertiary intrusion, the probable source of the minerals. Where fluids from the intrusions encountered the limestone and dolomite of surrounding rock, ore-bearing minerals were deposited. Though most mines in this district have played out, discovery of gold has now revitalized mining, at least for a time.

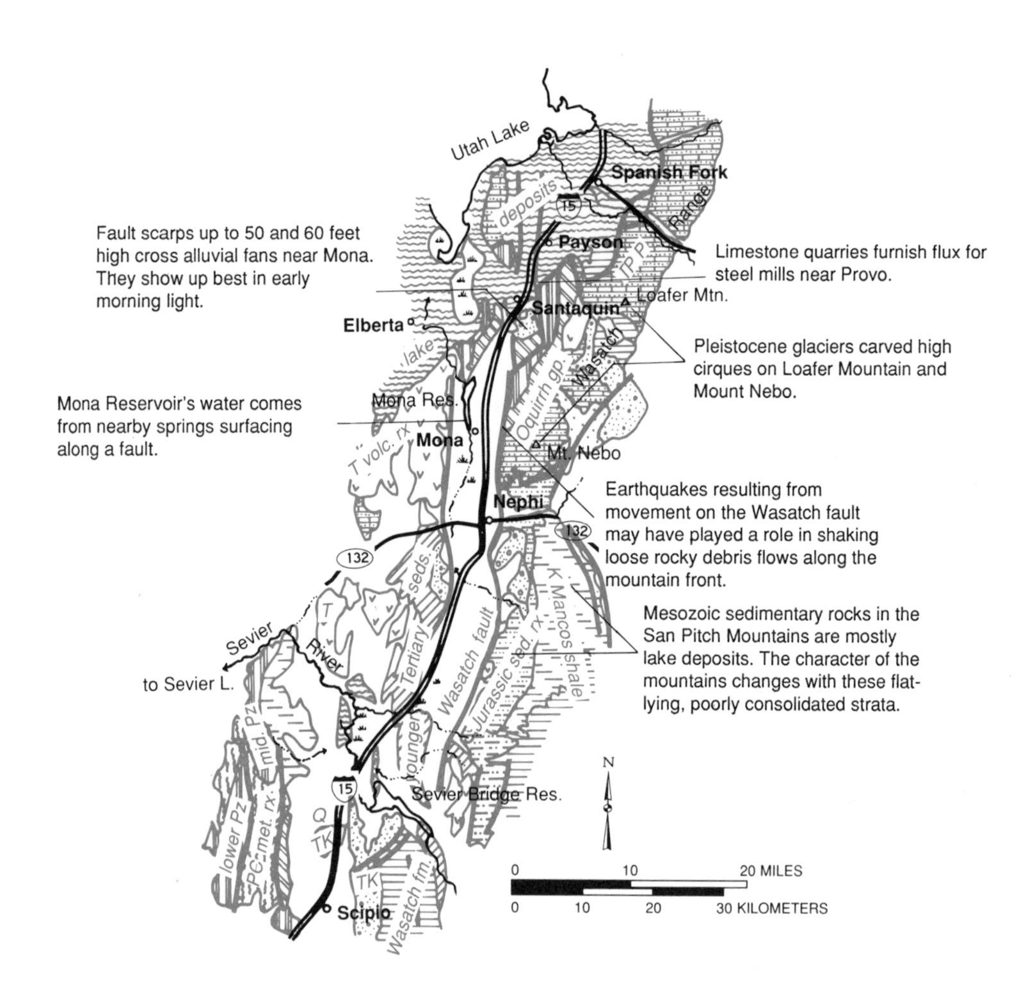

I-15
SPANISH FORK — SCIPIO

Near Santaquin the highway crosses a Pleistocene delta and a recent alluvial fan at the mouth of Santaquin Creek. Large quarries near exit 246 obtain Paleozoic limestone for use as flux in steel mills near Provo. Ground limestone added to smelter furnaces promotes melting of the ores and separation of the iron from remaining rock material.

At milepost 242 the highway crosses a large earthflow composed of rock and soil from the front of the mountain. A number of such features exist along the mountain front here; they can be recognized by their hummocky topography and coarse, bouldery debris embedded in a reddish, clayey matrix. Earthflows are very dense mudflows that carry lots of coarse rock material, even large boulders. Such flows result from unusually heavy rain; here they are also the product of steep mountain slopes and relatively weak rock on the mountain front. Lake Bonneville sediments lap onto the toe of the earthflow, so we know that it occurred before or during the time that lake existed. And the former existence of Lake Bonneville tells us that the ice ages were times of unusually heavy precipitation, which would favor the development of earthflows and landslides.

High on Mt. Nebo, three scoop-shaped cirques attest to the presence of alpine glaciers in Pleistocene time. Glacial meltwater drained into Lake Bonneville.

At the Utah-Juab County line the highway crosses a sand and gravel bar formed when Lake Bonneville was at its highest. Such bars exist around the entire perimeter of the former lake, along with other lakeshore features.

Three glacial cirques cut into the high summits of Mt. Nebo, east of Mona. The glaciers that carved them obviously clung with difficulty to the steep slopes of these mountains, spilling their rock debris down into stream canyons. There is not much in the way of glacial moraines except within the cirques. Glaciers extended down to about 9000 feet in this part of Utah, straightening and reshaping stream canyons down to that level. Below 9000 feet the unglaciated stream canyons are narrower than the glacially gouged valleys at higher elevations. Hummocky earthflows mark Mt. Nebo's lower slopes.

There are more earthflows south of Mona; see if you can spot them. Between mileposts 229 and 228, quarries visible at the base of the mountains produce road materials from the Wasatch fault zone. Between the quarries and Nephi, recent displacement along the Wasatch fault offsets alluvial fan surfaces by as much as 80 feet.

Near milepost 227, near Nephi, the Nebo thrust fault moved rocks of the Pennsylvanian and Permian Oquirrh formation southward over upside-down Triassic, Jurassic, and Cretaceous rocks. Rock relationships in this area are much more complex than can be shown at the scale of the maps in this book. The Nebo thrust, part of the Sevier thrust belt, extends northeast through the mountains as far as Heber Valley, where it is called the Charleston thrust fault, and east to Strawberry Reservoir, where it's known as the Strawberry thrust. All the mountains between here and there are made of the eastward-shoved upper plate of this great fault.

Light-colored rock being quarried near the mouth of Salt Creek Canyon at Nephi is the Jurassic Arapien shale, rich in gypsum.

South of Salt Creek Canyon the San Pitch Mountains, outposts of Utah's High Plateaus, contain lots of Cretaceous conglomerate and sandstone derived from the Sevier Mountains, which in Jurassic and Cretaceous time lay a short distance west of our highway route. The wedge of Cretaceous conglomerate, an alluvial apron much larger than the fans along the mountain front now, was in places 15,000 feet thick. Below the Cretaceous conglomerates are gypsum-rich Jurassic shales that weather into rounded gray-white hills. Nearer the highway, alluvial fans contain sand and gravel washed from the San Pitch Mountains. The highway climbs onto one of these fans at milepost 222. In the middle of the valley, fans from both directions meet.

Hills to the west between Nephi and Juab display silty limestones deposited in Tertiary lakes that lay in undrained basins between the newly risen Rocky Mountains and the newly formed tablelands of the Plateau country. Similar lake deposits make up the summit of the San Pitch Mountains near Levan. Much farther south, similar rocks have eroded into the intricately carved wonderland of Bryce Canyon National Park.

Between Levan and Scipio the country to the west opens up, giving us views of the Canyon Range with its long north-south strips of Precambrian and Devonian sedimentary rocks. The Precambrian rocks moved east over Devonian rocks along one of the faults of the Sevier thrust belt. On this side of the range the Precambrian rocks appear above Devonian strata; on the west side of the range they overlie Cambrian and Ordovician strata.

The Sevier River, halted in its path by the Sevier Bridge Reservoir, flows northward from the reservoir for several miles, meandering gently over the delta it built into Lake Bonneville in Pleistocene time — a quite extensive delta that reaches to and beyond the town of Delta. North of the Canyon Range, the river curves west and then southwest. For most of historic time it has lost itself in the sands and gravels of its old delta. But during the wet years of 1983-87 it once more reached the playa previously known and mapped as Sevier Dry Lake, creating a new lake, just a few feet deep, on the old playa surface. In Pleistocene time a slender arm of Lake Bonneville extended 25 to 30 miles up the valley of the Sevier River, toward Gunnison. Only the highest Bonneville shoreline is present there; the valley drained as the lake dropped from the Bonneville to the Provo level.

Near the Sevier Bridge Reservoir we leave the Wasatch fault, which continues south while we swing off to the southwest. Here we enter true Basin and Range country, with range after range coming into view to the west.

From mileposts 198 and 190 it is easy to see, on the Canyon Range to the west, the light-colored Precambrian quartzite that makes up the crest of the range, above stratified Devonian limestone and dolomite. The quartzite moved east as part of the Sevier thrust belt.

Scipio lies near the head of the long valley between the Canyon Range to the west, the Pavant Range to the south, and the Valley Mountains to the southeast.

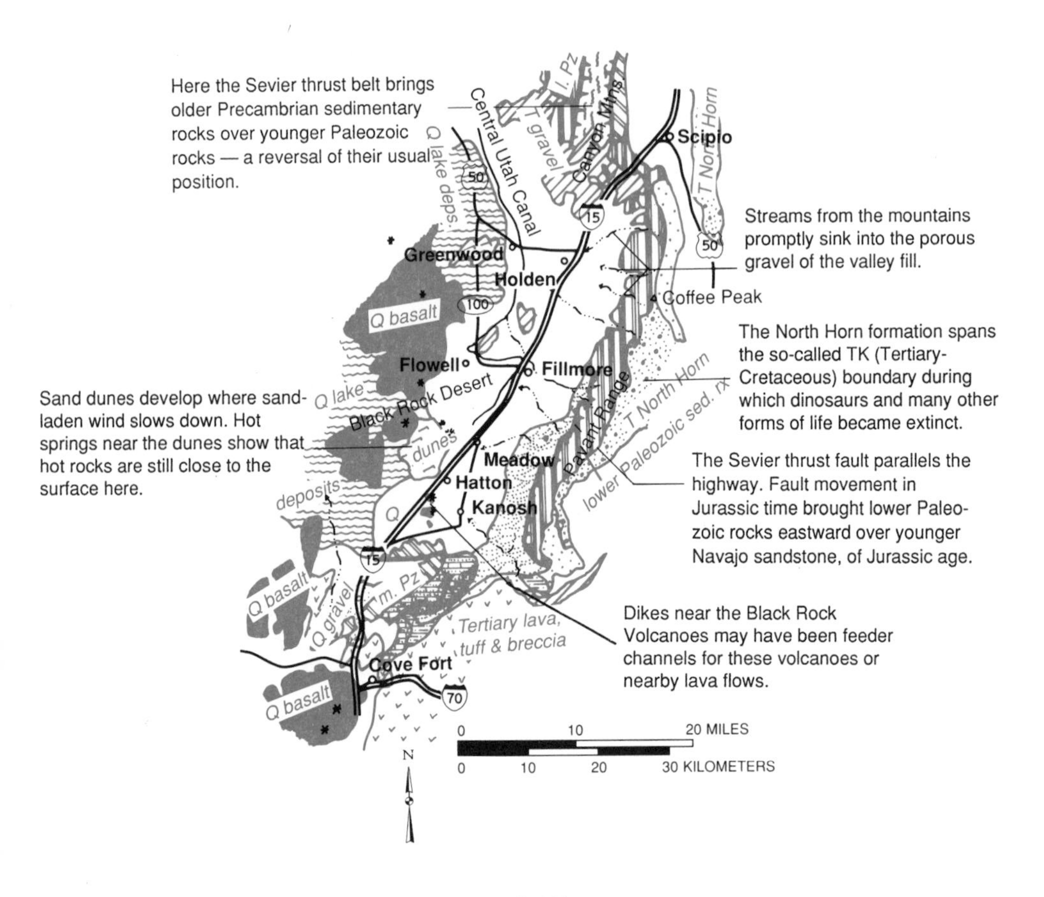

I-15
SCIPIO — COVE FORT

Interstate 15
Scipio—Cove Fort
55 miles/88 km.

The north-south ranges of the Great Basin meet the generally south-southeast trend of the mountains of central Utah at an angle, and so appear along this highway to be staggered, each offset from the next. The Wasatch fault is about 15 miles east of Scipio, on the other side of the Valley Mountains. The mountain front east of our highway is edged with other faults.

West of Scipio, in the Canyon Range, the northernmost of the staggered ranges we encounter, a thrust fault brings Precambrian quartzite, the light-colored cliffs and slopes at the top of the range, over gray Paleozoic rocks of the lower slopes. This thrust fault is part of the Sevier thrust belt, and the interstate more or less parallels its edge south to Cove Fort.

South of Scipio, high mountains east of the highway, in a more or less continuous line with the Canyon Range, are the Pavant Range. Thick layers of Cretaceous conglomerate top these mountains, above much-folded Lower Paleozoic strata. The massive Cretaceous rock, the Price River conglomerate, is part of a thick wedge of coarse sediment that developed as an alluvial apron east of the Sevier Mountains of Cretaceous time. It extended as far east as the Price River, north of the town of Price in eastern Utah.

Some of the Paleozoic sedimentary rocks can be seen as the highway climbs to Scipio Pass between the Canyon and Pavant ranges. These rocks were folded and distorted, as well as deeply eroded, before the conglomerate was deposited across them.

South of Scipio Pass the highway drops down into a broad, flat-floored valley, part of Lake Bonneville's lakebed, with the Pavant Range to the east. To the west, rising above the lake deposits, are lava flows centered around Pavant Butte, a Quaternary cinder cone scarcely eroded from its original form. Such cinder cones develop where basalt magma rising through the crust contains large amounts of steam. Escaping steam flings frothy blobs of lava from the volcanic vent to build the cone. Cinder cones generally form in the first stage of the eruptions, with lava flows pouring out through the porous, cindery base of the cones as or after the froth is expended.

Here, the cinder cones and their associated lava fields provide a name for the wide, flat-floored basin: the Black Rock Desert. In patches, these young basalt volcanic rocks extend south beyond Cove Fort, covering an area about 10 miles wide and 60 miles long. Like islands in a sea, many of the volcanic hills rise above the broad expanse of Lake Bonneville's former floor. Some of the lavas are believed to have erupted into the lake. Others are older — as shown by shorelines carved on some of the cinder cones, or younger — as shown by lavas overlapping lake sediments.

Though ranges surrounding the Black Rock Desert contain mainly Paleozoic strata, no such rocks underlie the Black Rock Desert or the Sevier Desert farther west. Instead, these deserts are brand-new land created in the last 20 million years or so by stretching and detachment faulting of the Earth's crust. Paleozoic rocks of the Cricket Mountains west of the Black Rock Desert used to connect directly with similar rocks in the Pavant Range. The gaping valleys opened by detachment faulting have largely filled with volcanic rocks, themselves a result of the faulting, and with sediments washed from adjacent mountains.

The entire volcanic area is a geothermal region. There are hot springs west of Holden, near Hatton, and elsewhere on the Black Rock Desert. Heat sources related to the volcanism are still quite near the surface here; both volcanism and hot springs are products of the active faulting along the east edge of the Basin and Range region.

Near Holden, the Price River conglomerate caps the Pavant Range to the east, with Cambrian rocks below it. The edge of the Sevier thrust belt is just west of us, under the lake deposits. The younger Basin and Range faults cut across the older Sevier thrust fault.

Near Fillmore, several cinder cones have been quarried for lightweight aggregate, the bubble-filled volcanic cinders seeming ready-made for such use. Several buttes of Tertiary sedimentary rock protrude above the lake floor near Fillmore.

From Fillmore south, the Pavant Range contains Cambrian strata thrust across much younger Mesozoic strata. The light-colored lower slopes are Navajo sandstone, a massive, dune-formed Jurassic sandstone that creates much of the scenery of the Plateau country, equivalent of the Nugget sandstone of northeast Utah. The highest peak on these mountains, White Pine Peak (10,270 feet), consists of strata that are thought to span the boundary between Cretaceous and Tertiary time, the North Horn formation. Rocks younger than Cambrian and older than Cretaceous were eroded away before the North Horn formation was deposited. Their absence here is part of the evidence for the Sevier mountain building event, which raised this area so that it was subjected to intense erosion.

West of Meadow and Hatton the black lavas of the Black Rock Desert contrast sharply with nearby dunes of white gypsum sand. The gypsum comes from highly saturated springs in this area. Low, light brown hills directly west of Hatton are tufa cones, spongy deposits of calcium carbonate, near Hatton Hot Springs. Such cones form around hot springs as the water cools and evaporates, precipitating some of the calcium carbonate it carried in solution.

Near milepost 153, a short distance south of Hatton, the highway passes between the two Black Rock Volcanoes. Reddish soils in this area are derived from the volcanic rocks. Twin Peaks, small light-colored peaks rising above the southern end of the Black Rock Desert, are lava domes about three million years old.

The highway crosses a pass here, climbing a gravel terrace dating back to Lake Bonneville days, a stream-formed delta at the base of the mountains. Slices of crushed, folded, broken, and in some cases completely overturned Paleozoic sedimentary rocks border the highway on the approach to the pass. These Paleozoic rocks are part of the Sevier thrust belt, crumpled and broken in the process of being pushed or pulled far east of their original position. The edge of the thrust is below the summit rest area, hidden by some volcanic rocks. These

Old Cove Fort was built of volcanic rocks: Quaternary basalt for the walls, and softer, more easily worked Tertiary tuff for the fireplaces and chimneys. The basalt was quarried near the junction of I-15 and I-70.

lighter-colored volcanic rocks are older and quite different from the black lavas and purple cinders we've been seeing: They are pinkish volcanic tuff, mudflow deposits, and breccia, silicic rather than basaltic, that date back to explosive eruptions of Tertiary volcanoes.

South of the pass are views of another recently active volcanic region, with a small flat-topped cinder cone rising above the lava flows of a small shield volcano.

Cove Fort, too, is in a geothermal area, with nearby hot springs. Sulfur deposits in this area, mined for a time, are associated with some of the volcanic rocks and hot springs. Cove Creek, which flows through the town, divides the Pavant Range to the northeast from the Tushar Mountains to the southeast.

Interstate 15
Cove Fort—Cedar City
76 miles/122 km.

Just south of the Cove Fort interchange, the highway passes two cinder cones rising atop lava flows west of the highway, one of them quite near the interchange at milepost 132. As described in the previous section, cinder cones develop when molten basalt magma interacts with groundwater to froth into popcornlike cinders, which are blown from volcanic vents and pile up around them.

To the east is a much larger volcanic region centered on Mt. Belknap, also including Signal, Delano, City Creek, and Baldy peaks. Lifted by Basin and Range faulting about 10 million years ago, these mountains now form the highest range between the Rocky Mountains of Colorado and the Sierra Nevada far to the west.

The volcanic materials of these mountains are lighter in color than basalt, and silicic in composition. Eruptions of silicic volcanoes commonly come in the form of huge explosions of volcanic ash followed by quieter emission of thick, sticky lava that bulges up as volcanic domes, the type of eruption seen in Mt. St. Helens during and after 1980. This part of Utah saw many such explosions in mid-Tertiary time, as is testified by the unbelievable deposits of tuff in this and adjoining areas. Some of the tuff can be identified as coming from other volcanic regions such as Yellowstone and parts of California. The volcanic violence of Miocene time far exceeded any known in historic time.

Visible to the southwest from milepost 130 and other vantage points is evidence of another type of igneous activity: the large granite intrusion of the Mineral Mountains. Even though their core is not metamorphic rocks, these mountains are considered to be a metamorphic core complex range that formed in Miocene time. Soon after the intrusion had cooled and hardened, rock layers above it slid west along a detachment fault. As the weight of overlying sediments was thus reduced, the granite core of the mountains bobbed up, so to speak, to form the range. The detachment fault involved here may be part of the large Sevier Desert detachment fault involved in creation of the Sevier and Black Rock deserts.

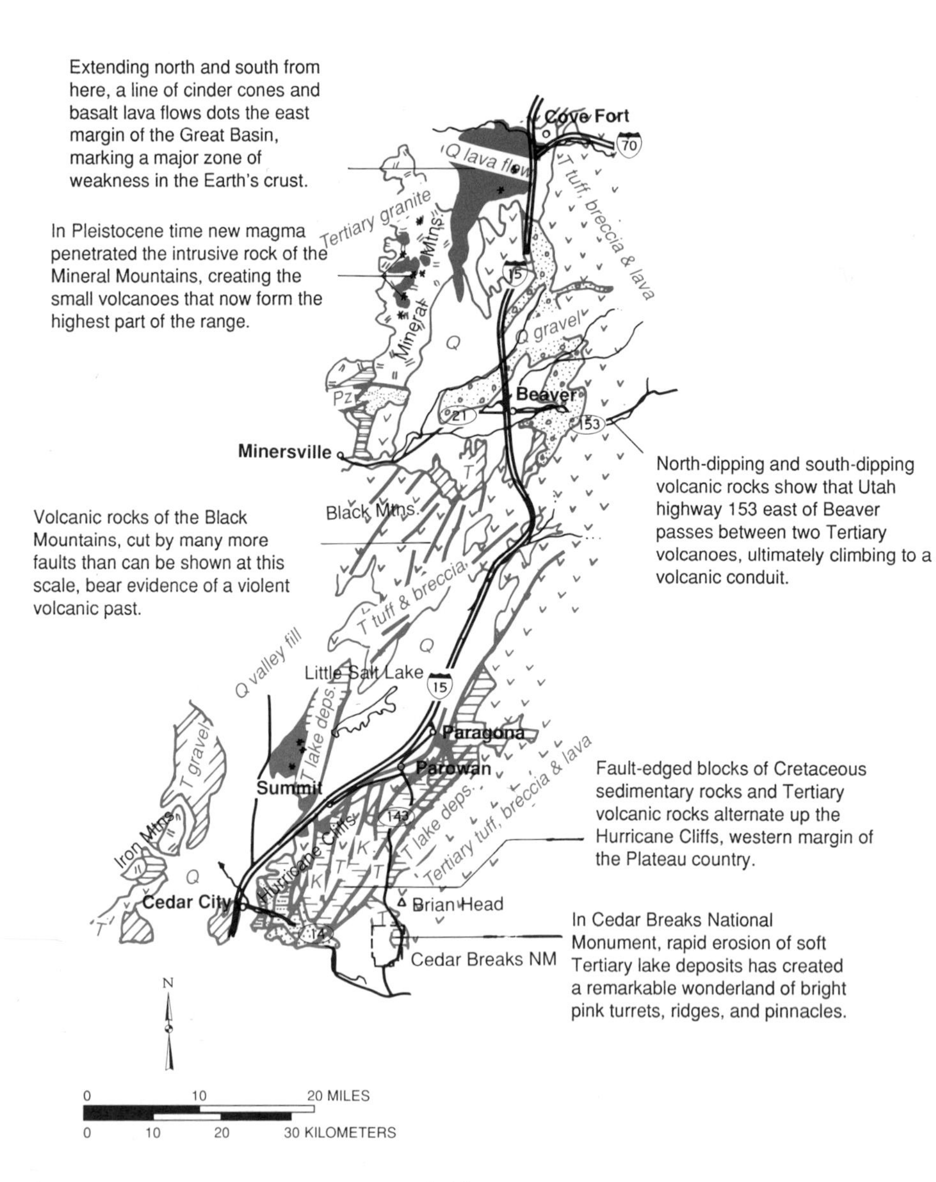

I-15
COVE FORT— CEDAR CITY

A row of small volcanoes, formed long after the intrusion, dots the long summit of the Mineral Mountains. Despite their name, these mountains have not been very important in Utah's mining history. A few mines cluster near the south end of the range, near the town of Minersville, where mineral-rich fluids from the intrusion came in contact with Paleozoic limestone. In such areas, limestone acts as a catalyst, helping to precipitate ore minerals from fluids accompanying the intrusion.

As the highway approaches milepost 126, it climbs onto a terrace of Pleistocene gravel deposited by heavily burdened rivers of ice-age time. The terrace is cut by many small normal faults, an indication that fault movement is still going on here.

From the top of the divide between mileposts 119 and 118, a hilly region of much-faulted Tertiary volcanic rocks can be seen to the south. This range, the Black Mountains, extends well to the west beyond the Mineral Mountains. South of it lies the Markagunt Plateau, fronted by the Hurricane Cliffs. There, flat-lying Mesozoic rocks are capped with still more Tertiary volcanic rocks. Quite a sizeable volcanic region. Think of the immense amount of volcanic material emptied from all these volcanoes! Depending on the spacing

Utah 153 leads into the Tushar Mountains to the conduit of one of the volcanoes of which these mountains are composed. Harder than surrounding layers of lava, breccia, and volcanic ash, igneous rock frozen in the conduit resists erosion.

of the eruptions, it is likely that clouds of ash, carried by stratospheric winds, markedly affected the Earth's climate in Tertiary time.

The highway crosses two more Pleistocene terraces before reaching the town of Beaver. The Beaver River, joined by Wildcat and Indian creeks and other tributaries, nourishes this fertile valley. An interesting side-trip up Utah 153 east of Beaver leads into the heart of the Tushar Mountains. By watching the dip of the volcanic layers, which is the original dip at which they were deposited on the sloping sides of their volcanoes, you can discover that part of the route runs right between two volcanoes. Ultimately the road climbs steeply up to the very heart of a volcano whose still recognizable conduit has been bared by erosion.

South of Beaver, the highway crosses the east end of the Black Mountains. The range's dense concentration of faults, most of them trending southwest to northeast, continues east into the low area between the Tushar Mountains and the Markagunt Plateau. Some of the faults displace the surface of late Tertiary gravels north of the range, and so must have seen movement in Quaternary time, the last two million years.

The Parowan Valley, south of milepost 95, is another flat-floored valley once occupied by a lake. Isolated by ridges to the west and south, however, this was not part of Lake Bonneville. It is now the site of Little Salt Lake, an ephemeral pond, a few miles long, that dries up completely in dry years. The valley is closed at its southern end by Quaternary lava flows and fault blocks of Cretaceous and early Tertiary sedimentary rocks. To the east are more Tertiary volcanic rocks.

Near Paragonah and Parowan the fault slices to the east include lake deposits of the Claron formation, an early Tertiary deposit that forms the sculptured wonderland of Cedar Breaks National Monument and Bryce Canyon National Park. Here the mountain front rises abruptly along the Hurricane fault. More than 160 miles long, this fault extends south into Arizona, dividing the Plateau country from the Basin and Range region to the west. Displacement along it amounts to several thousand feet, and creates the Hurricane Cliffs.

From the milepost 83 rest area, sculptured red rocks of the Claron formation can be seen to the east. Soft, easily eroded, with limy siltstone layers alternating with silty limestones, this formation erodes into several breaks along the margin of the Markagunt Plateau. In Bryce Canyon National Park farther east, the same formation is a paler pink.

The interstate remains close to the Hurricane fault and the Hurricane Cliffs as it continues south to Cedar City. The mountainous area west of Cedar City includes many mineral deposits, and mines to go with them. Iron, silver, gold, lead, and zinc have been produced from this region, with ores occurring near small Tertiary intrusions. The intrusions domed up some of the sedimentary rocks which lay above them, in this case Cretaceous rocks that now appear in a ring around the base of the mountains. In the Escalante Desert west of the mountains is the southernmost evidence of Lake Bonneville: lake sediments and shorelines more than 220 miles south of today's Great Salt Lake.

Coal and uranium have been mined from Cretaceous rocks on the Markagunt Plateau east of Cedar City.

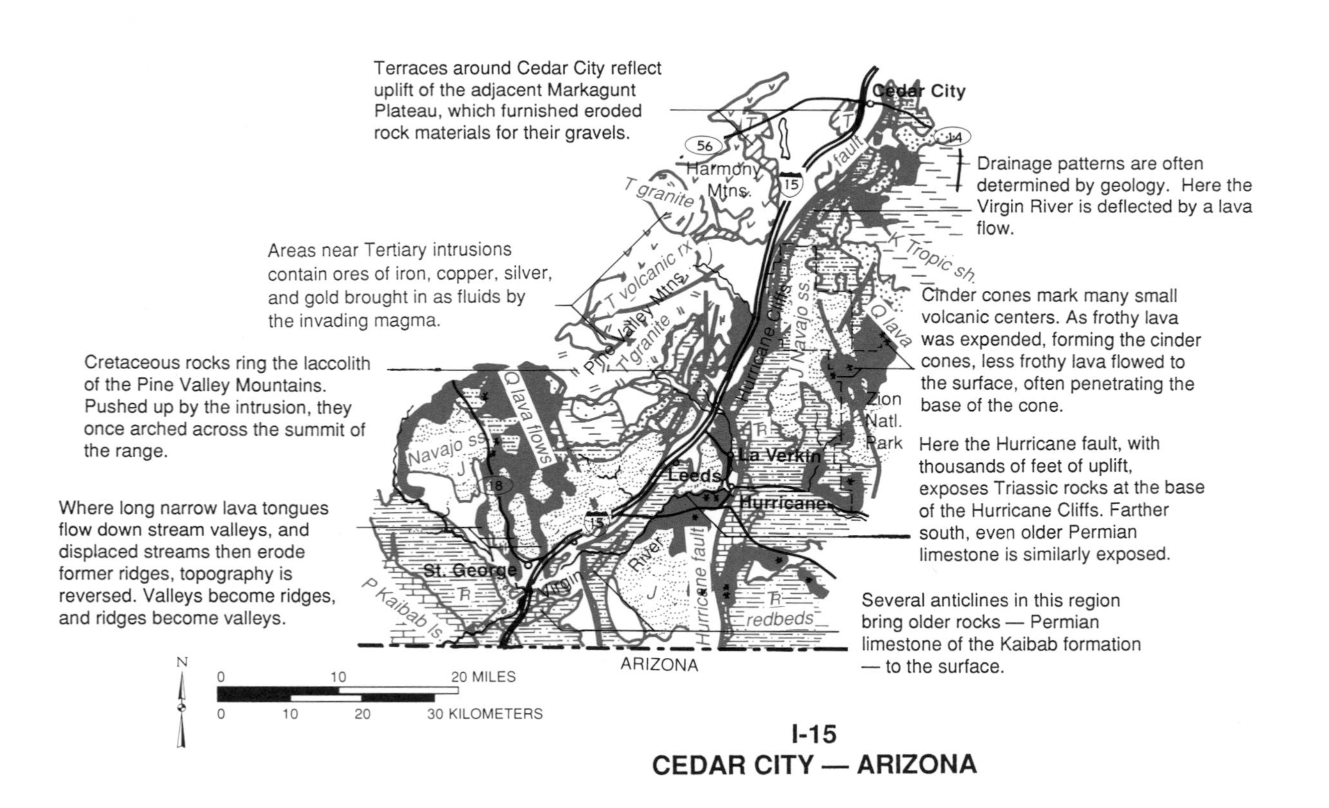

I-15
CEDAR CITY — ARIZONA

Interstate 15
Cedar City—Arizona
61 miles/98 km.

West of Cedar City, Tertiary intrusions form the low hills of the Iron Springs Mining District. Here as in many other parts of Utah, ores result from interaction between Tertiary intrusions and surrounding rocks, particularly Paleozoic and Mesozoic limestone. The intruding magma domed Jurassic and Cretaceous strata and reacted with Jurassic limestone, creating deposits of magnetite and hematite, ores of iron.

East of Cedar City, along the Hurricane Cliffs, colorful Jurassic and Triassic rocks include the Navajo sandstone, the prominent white cliff-former visible on the mountain front. The cliffs mark the line of the Hurricane fault, the major fault separating the Colorado Plateau and its subdivision the Markagunt Plateau from the Basin and Range country west of it.

Cretaceous rocks lie more or less horizontally above the Jurassic and Triassic strata: the Dakota sandstone and Tropic shale in forested slopes, and the Straight Cliffs sandstone capping the escarpment. From high spots on the highway west of town you can glimpse younger rocks: pink Tertiary lake deposits, eroded in one of several breaks that edge the Markagunt Plateau. The succession of cliffs is that found elsewhere in the Plateau country: from bottom to top the Vermilion Cliffs of Triassic rocks, White Cliffs of the Jurassic Navajo sandstone, the Cretaceous Gray Cliffs, and the Pink Cliffs of Tertiary rocks. (A stratigraphic section of these rocks appears on p. 25.) Here and there dark gray basalt forms the plateau's surface or pours down over its edge.

South of Cedar City, between mileposts 55 and 54, the highway cuts through a lava flow ridge, exposing the baked soil beneath the flow, reddened like brick by heat from the hot lava. Small faults cut through and displace parts of the flow.

More details of the edge of the Plateau country show up from the rest area south of milepost 45. Prominent cliffs to the east are the Kolob Fingers, massive ridges of Navajo sandstone extending west from the Kolob Terrace. They are part of Zion National Park, beautiful but less well known than Zion Canyon itself. Colorful Triassic rocks appear below the Navajo sandstone, and some whitish Permian limestone below them. In places, slivers of red Triassic siltstone are faulted downward against the limestone.

More basalt lava flows appear on Black Ridge, near exit 36. As is commonly the case with lava flows, these flowed down a valley, displacing the stream. Less resistant rocks on either side of the valley then eroded away, leaving the hard lava as a ridge. Near its center the lava is quite thick, as you can see in the deep gorge at milepost 37.

The Pine Valley Mountains to the west consist of a large laccolith, another intrusion that in Tertiary time pushed upward here, breaking through the lower sedimentary rock layers, then spreading out beneath those above. On the map, the intrusion is a queer shape, cut off on the north by a fault. There are also Miocene volcanic rocks, tuff and breccia, in the Pine Valley Mountains, products of the great surge in volcanic activity associated with the detachment faulting that shaped the Basin and Range region.

The breakaway zone related to the detachment faulting is thought to lie along the west side of the Pine Valley Mountains. The High Plateaus rose buoyantly as the rocks that once covered them slid west on the detachment fault.

Dark basalt across the valley north of Pintura is Black Ridge. It is the same age and composition as that on top of the Hurricane Cliffs to the east. Geologists believe they are parts of the same lava flow, less than two million years old. The difference in their elevation is therefore a measure of displacement on the Hurricane fault within the last two million years.

Permian marine limestone, the Kaibab limestone, appears low down on the Hurricane Cliffs near Pintura. The Hurricane fault is at the base of the light-colored rocks, which include the Kaibab limestone, the Toroweap formation below it, and the Coconino sandstone below that. These three formations, all Permian, record the change from land to sea in Permian time. The Coconino sandstone was deposited in Permian sand dunes; the Toroweap formation is a near-shore marine deposit recording a single incursion and withdrawal of the sea, and the Kaibab limestone developed during a later and more extensive marine incursion. These rocks are in a narrow fault sliver between two branches of the Hurricane fault.

West of the highway is the boulder-dotted surface of a mudflow that came from the Pine Valley Mountains.

Most of the many small faults here trend parallel to the Hurricane fault. The majority, including the Hurricane fault, are fault zones rather than simple breaks. Small anticlines and synclines parallel the faults as well. Evidence of volcanism, in the form of cinder cones and basalt lava flows, seems to concentrate near the Hurricane fault.

South of Pintura our highway swings southwestward, away from the north-south Hurricane fault.

Leeds is in the heart of the Silver Reef Mining District, with old silver mines to the west and uranium mines in Triassic rocks near and east of the town. Both silver and copper minerals were unusual: The ores replaced fossil leaves and other organic material in the Jurassic Silver Reef sandstone, part of the Moenave formation. A trio of similar sandstones cap cuestas, long narrow hills formed on tilted rock layers, in this area. Except for these moderately resistant sandstone layers, the Chinle formation is soft and shaly, eroding into bluish gray badlands. In places, it contains fossil wood.

Between mileposts 20 and 19, the highway runs along a valley eroded in tilted layers of soft, shaly mudstone and siltstone, between cuestas capped by harder sandstone layers of the Chinle formation. Coming out of this valley watch roadcuts for good exposures of the soft and hard layers. Silver Reef mine dumps can be seen here and there to the west.

Highway cuts farther southwest contain the coarse, cobbly gravel of an old alluvial fan, with pebbles and cobbles in large part derived from the Pine Valley Mountains to the northwest.

Between mileposts 16 and 15 another inverted valley shows up as a long ridge capped by a lava flow that fills a former valley of the Virgin River. The displaced river and one of its tributaries cut new valleys in soft sedimentary rocks on either side, leaving the lava flow ridge. The basalt contains vertical columnar joints that developed as the lava crystallized.

A roadcut between mileposts 18 and 17 exposes the cobbly gravel of another old stream course, one that wasn't filled with lava. And near Washington is another that was. Lots of faults in this area complicate the geologic picture, tilting layers of rock that were deposited in horizontal position. Basically, the sedimentary rocks swing in a horseshoe around the southern end of the Pine Valley Mountains. Between Leeds and Washington, just southeast of the highway, they bend upward into the long, narrow Virgin anticline. Utah 15 to Zion crosses this fold in its first few miles.

Except for the lava flows, all the rocks near St. George are Triassic and Jurassic sedimentary formations. Each has its characteristic color, mostly tones of red and pink, and way of weathering. Most of the dark red siltstone and shale are river floodplain and delta deposits. Paler, cross-bedded sandstone generally signifies flowing streams or wave-washed bars or sand dunes. The massive, light-colored Navajo sandstone that appears on the southern slope of the Pine Valley Mountains accumulated as dunes in the vast sand sea of Jurassic time.

Hemmed in by lava flows, St. George occupies a flat-floored valley that contrasts with all the surrounding ruggedness. The town lies on Chinle formation shales that in some places contain petrified logs. Reddish cliffs just north of town are eroded in Jurassic rocks. The Hurricane Cliffs rise about 12 miles away to the east, along the Hurricane fault. Long, fingerlike lava flows approach the city from the north, again filling old stream channels on the flanks of the Pine Valley Mountains. Drab badlands of soft Chinle shales, rich in volcanic ash and uranium, form playgrounds for off-road vehicles. Through it all, south of town, the Virgin River, born in the highlands of the Markagunt Plateau, flows toward Lake Mead and the Colorado River, swinging in widened meanders where its gradient is low. Crossing the river's floodplain, the highway rises onto red Triassic siltstone and mudstone as it approaches the Arizona line.

The Beaver Dam Mountains west and southwest of St. George are one of Utah's metamorphic core complex ranges. Their surface is part of the detachment fault surface that underlies the western deserts; their western flank exposes a portion of the crust unroofed by movement on the fault.

The Bonneville salt flats are so flat and level that they reflect the curvature of the Earth's surface. From a standing position, a tall man can see about three miles — not enough to see both ends of the Bonneville Speedway at the same time.

Interstate 80
Nevada—Salt Lake City
122 miles/196 km.

Lake Bonneville, the largest freshwater lake of ice-age Utah, extended west well into Nevada, its western shore in places 10 miles beyond the Utah line. Where Interstate 80 crosses the state line near Wendover, recognizable Lake Bonneville shorelines crease the lower slopes of the Silver Island Mountains to the north. This range consists of steeply tilted Paleozoic rocks, and is one of the many fault-block ranges of the Basin and Range Province. As elsewhere in this region, faults that edge the mountain block are hidden beneath the basin sediments.

Lake deposits floor the desert east of these mountains, and show us that not long ago, geologically speaking, the salty ancestor of today's Great Salt Lake also extended this far west. The salt is not just table salt, but includes gypsum, potash, calcium carbonate, and other minerals carried in solution by streams entering the lake from the mountains of northern and northeastern Utah. The various chemicals were then concentrated by evaporation of lake waters. A chemical plant northeast of Wendover uses some of these deposits, producing potash for fertilizer.

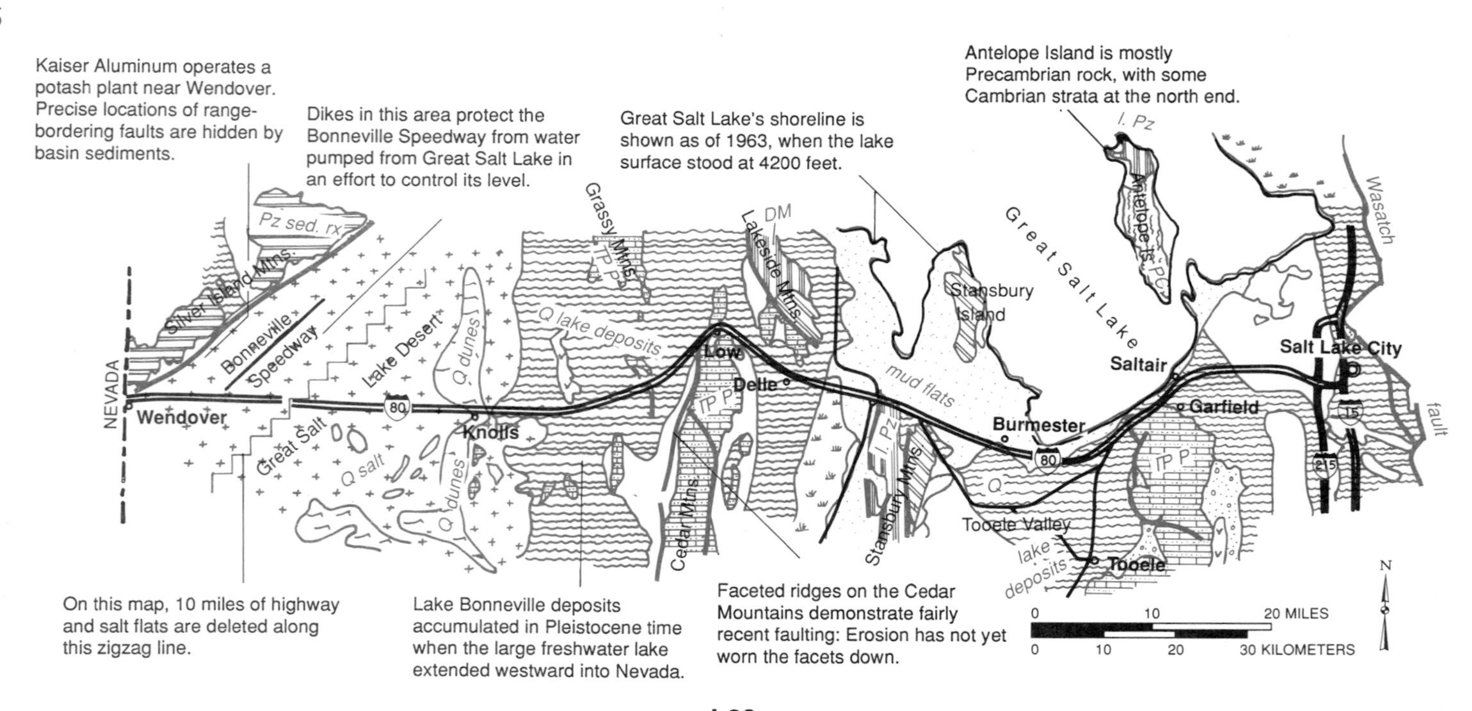

I-80
NEVADA — SALT LAKE CITY

On the smooth, hard-packed surface of the salt flats, virtually devoid of vegetation, is one of the famous racetracks of the world, the Bonneville Speedway, site of numerous world speed and endurance records. The speedway, about 12 miles long and 80 feet wide, is so flat that it curves, a seeming contradiction in terms, with the Earth's curvature. Without climbing on some kind of platform, you can't see both ends of it at once! This curvature can be seen quite well from the rest area east of Wendover, where an exhibit gives details of racetrack history. The 1983-87 rise in the level of Great Salt Lake, and the necessity of pumping water from the lake into desert basins farther west, endangered the Bonneville racetrack; long dikes built to hold back the water can be seen from the highway.

The interstate crosses about 40 miles of salt flats here. You can ask for no more barren desert than this. Around the few plants able to tolerate this much salt, wind has scoured away some of the salt, leaving the plants on low pedestals where their roots hold salty soil together.

Scattered sand dunes appear north and south of the highway near milepost 39. Gypsum from the salt flats is a major component of these dunes. Larger dunes have built up near Knolls, where prevailing westerly winds, bouncing the grains of sand close to the surface, are slowed down and forced to rise as they approach the Cedar and Lakeside mountains. Some outlying ridges of Paleozoic sedimentary rocks appear near Knolls, with dunes climbing their western slopes.

East of Knolls the land rises, and we climb almost imperceptibly above the level of the salt flats and onto older Lake Bonneville deposits. These are mostly fine silt, without salt because Lake Bonneville was a freshwater lake.

Both the Cedar Mountains and the Lakeside Mountains are typical of the fault block mountains of the Basin and Range region. The Cedar Mountains consist of east-dipping Pennsylvanian and Permian sedimentary rocks, mostly limestone and sandstone, that belong to the Oquirrh group. Rocks in the Lakeside Mountains north of the highway are older, Cambrian to Mississippian, with the strata bent into a syncline. Both are tilted blocks whose marginal faults flatten with depth, curving and perhaps merging several miles below the surface with the detachment faults of this part of Utah. Many fossils can be found in these ranges. The horizontal shelves of old Lake Bonneville shorelines mark both ranges, but shorelines show up best on smaller buttes, those with minimal drainage areas and therefore minimal stream erosion.

Light-colored Cambrian quartzite at the core of an upward-faulted anticline forms the central ridge of the Stansbury Mountains.

From the top of the overpass at exit 76, Great Salt Lake appears to the east and north, with the Stansbury Mountains south of it and Stansbury Island out in the lake. At low lake levels mudflats connect the island to the shore.

The Stansbury Mountains and Stansbury Island are parts of a long, narrow fault block in which Paleozoic rocks, in addition to being faulted upward, arch into an anticline. The anticline can be seen on the south end of Stansbury Island, a continuation of the structure shown in the Stansbury Mountains. The island is part of one of the raised fault blocks that separate the three long north-south basins flooded by Great Salt Lake.

East of the Stansbury Mountains the highway crosses Tooele Valley, once an arm of Lake Bonneville. Notice the flatness of the valley floor, characteristic of basins filled with fine lake sediments. A fine example of the horizontal nature of water-laid deposits. Close to Great Salt Lake, numerous manmade salt pans employ western Utah's plentiful solar energy to concentrate lake brine. Rising lake levels in the mid-1980s flooded many of these pans, forcing salt companies to close their plants, temporarily at least. Both railway and highway were raised several feet to keep them above the rising waters.

As we cross Tooele Valley south of Great Salt Lake there are good views ahead of the Oquirrh Mountains. The north end of this range consists almost entirely of Pennsylvanian and Permian sedimentary rocks of the Oquirrh group, part of Utah's complexly folded and faulted Sevier thrust belt. The south end includes several intrusions of Tertiary granite, one of them responsible for the rich copper ore at

Bingham Copper Mine. The mine itself, the world's largest open pit copper mine, is on the east side of the range. Its ores contain molybdenum, gold, lead, and zinc as well as copper. They are processed at the smelter near I-80; the large tailings pond can be seen east of Great Salt Lake State Park.

Prominent Lake Bonneville shorelines, some of them with porous white lake deposits called tufa, crease the slopes of the Oquirrh Mountains. Similar shorelines can be seen on the Wasatch Range beyond Salt Lake City, and on Antelope Island north of Saltair and Great Salt Lake State Park.

Antelope Island is another mountain ridge faulted along both edges, with lake-filled grabens on either side. Tiny Fremont Island north of it, and the Promontory Mountains jutting into the northern part of Great Salt Lake make up the second raised fault block ridge that divides the lake's basins. Rocks visible on the south end of Antelope Island are Precambrian gneiss and schist, some of the oldest rocks in Utah, adjacent to some of the youngest sediments — those forming in the lake today.

Much of Salt Lake City lies on deltas built out into Lake Bonneville by streams draining the Wasatch Range: City, Emigration, Parleys, Mill, and Big and Little Cottonwood creeks. During the ice ages, these streams carried abundant meltwater from glaciers in high mountain valleys. The western part of the city spreads onto the alluvial plain of the Jordan River, which drains Utah Lake, a freshwater lake that like Great Salt Lake occupies only part of former Lake Bonneville's immense basin.

Wave-cut shorelines are prominent features on slopes surrounding Great Salt Lake. In 1890 geologist G.K. Gilbert recognized them as the work of a much larger lake of ice-age time. Here, the shorelines cut across contorted strata of the Oquirrh formation, part of the Sevier thrust belt.

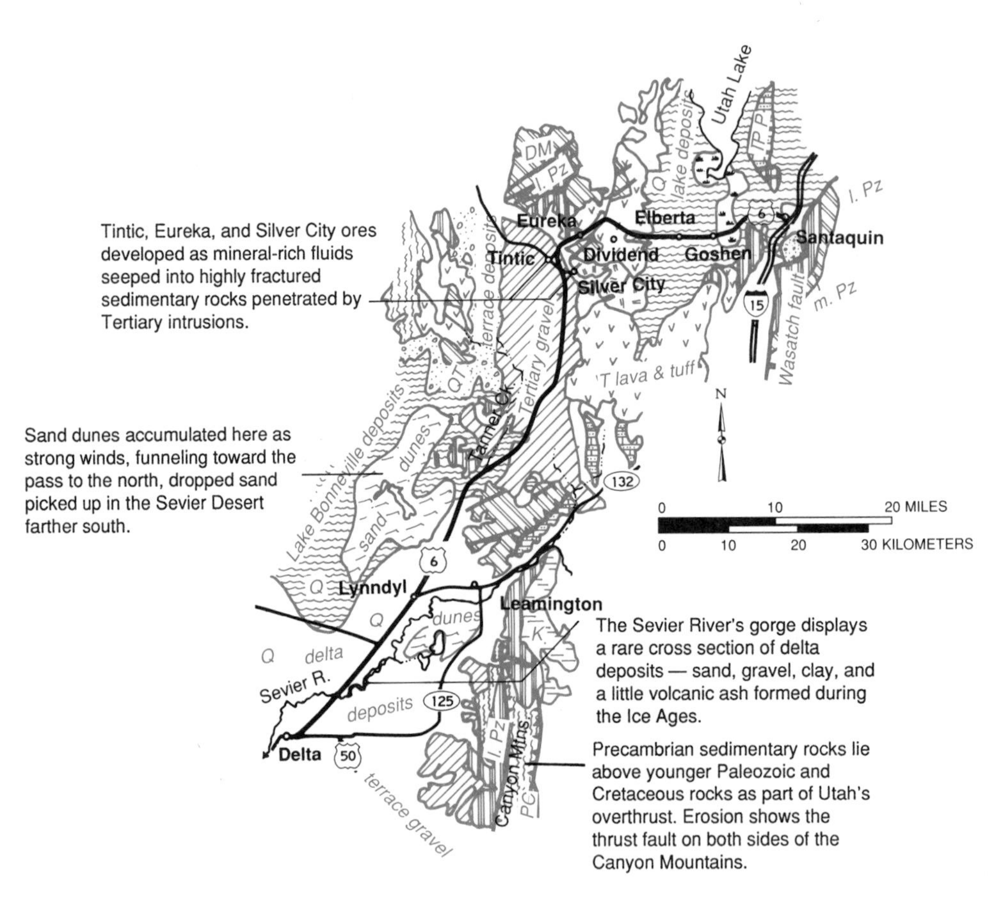

US 6
SANTAQUIN — DELTA

US 6
Santaquin—Delta
69 miles/111 km.

Leaving Santaquin on the wide Provo shoreline of ancient Lake Bonneville, US 6 soon drops down onto a level surface of non-delta lake deposits. Between the highway and Utah Lake is West Mountain, whose Quigley Quarry provides limestone for smelters at Orem.

Salt surfacing the white flats near Goshen is deposited by small salt springs farther west. Every year, as spring waters evaporate in the hot summer air, their minerals remain, resurfacing older salt layers.

Deep gullying of the draw at milepost 151 is a fairly recent development; it exposes in cross section some of the fine lake deposits that make up the valley floor. Ever since its separation from its big brother to the north, Utah Lake's water has been fresh; an outlet via the Jordan River, flowing into Great Salt Lake, prevents salt buildup. Most of the water comes from the Provo River, with headwaters in the Uinta Mountains of northeast Utah.

West of Elberta the highway climbs steadily across a broad alluvial fan dotted with boulders from the East Tintic Mountains, the small range to the west. This range is dotted with old lead and silver mines, most of them now idle. The area is rich in ores of silver, gold, copper, iron, and other metals. The West Tintic Mining District a few miles

Old mines and miners' cabins dot the hills around Dividend, Eureka, and Tintic. The East Tintic Mining District was once the bustling silver capital of Utah.

farther west was also a mining area. Tertiary intrusions brought in the ore-bearing solutions, which solidified in cracks and fissures in both the intrusions and surrounding Paleozoic limestone. Many ore bodies discovered here were cut off at the top by Tertiary volcanic rocks.

Outcrops of Tertiary granite can be seen in the Tintic area, as near milepost 144. In places, the granite has rotted into coarse sand colored yellowish or reddish brown by oxidized iron minerals that coat mineral surfaces. Early prospectors searched for areas with just such signs of mineralization, and in this area they struck a bonanza. The region contains some of the richest and most productive mines in United States: 18.4 million tons of ore have been mined here, with these production figures for the final products:

gold	2.8 million ounces
silver	275 million ounces
copper	125,000 tons
lead	1,160,000 tons
zinc	216,000 tons

Eureka is another mining town. Several of its mine dumps are now being reworked to extract minerals left behind by older, less efficient extraction methods. Most of the mines are in Cambrian and Ordovician limestone. Many old workings can be spotted by their dumps and headframes in the hills near town. Headframes served to support sheaves — large pullylike wheels — for the cables by which elevator platforms and ore buckets were raised and lowered in the mine shafts. Dumps are predominantly low-grade waste rock removed while digging toward usable ore. But the old mine dumps are good places for mineral hunting. Most are on private property, so get permission before you start looking for specimens. Stay away from open shafts, adits, and areas where the ground is cracked or collapsed.

Other mines here are glory holes, cone-shaped cave-ins that opened as large masses of ore were removed from below.

At Tintic Junction west of the mountains the highway turns south down the broad valley of Tanner Creek, between the East and West Tintic Mountains. The valley is floored with Late Tertiary sandstone, mudstone, siltstone, volcanic ash, and salt, in places thousands of feet thick. Lake Bonneville shorelines mark the mountain slopes.

A short distance south of Silver City junction, dumps east of the highway mark a mine producing a fine white clay called halloysite.

When this mine was operative, cables ran from hoisting engines in a nearby building up over the sheaves in the headframe, and then down into the mine, to lower and lift men, mules, and ore buckets.

The highway continues southwest into the Sevier Desert, a broad, open basin ringed by mountains. It is easy to see that this valley was part of Lake Bonneville: Shorelines mark most of the surrounding slopes. At the northeast end of the valley is a large dune area known as the Little Sahara or Lynndyl Sand Dunes. Dunes develop where there is a source of sand and plenty of wind. The sand comes from Pleistocene delta deposits of the Sevier River, which we will see farther south. Southwest winds flow freely across the Sevier Desert, picking up the sand. Sand Mountain, in the midst of the dune field, apparently slows the wind, causing it to drop its load of sand. Dunes of several characteristic shapes occur here: low parabolic dunes trailing long arms stabilized by vegetation, crescent-shaped barchan dunes, transverse dunes formed of coalescing barchan dunes, and climbing dunes on the flanks of Sand Mountain. Spreading vegetation has stabilized many of the dunes, especially those around the margins of the dune area.

East of US 6 a low fault scarp marks the surface of alluvial fans sloping gently from the mountains.

Near Leamington the Sevier River slips through a narrow slot between the Gilson Mountains and the Canyon Mountains, and enters the Sevier Desert. During the ice ages this river, laden with rock debris from glacier-capped mountains of central Utah, built a delta more than 20 miles across out into Lake Bonneville. It extends to the town of Delta and beyond, dividing the Sevier Desert into two broad basins.

Entrenched in its former delta, the Sevier River of historic time ran only after heavy rains — partly Nature's doing, with this dry, porous delta to sink into, and partly man's doing, with its upstream waters diverted for irrigation. With the heavy rains of the 1980s, the river flowed for a time. Its gorge is east of the highway from mileposts 103 to 94, where the highway crosses it. Delta deposits, exposed in bluffs near the river as well as in highway cuts, consist of sand, silt, gravel, and a thin layer of black volcanic ash. The river's downward cutting, through its former delta, provided some of the geologic information used in the reconstruction of Lake Bonneville's up-and-down history.

A plant a few miles south of Lynndyl processes beryllium, a strong, lightweight, heat-resistant metal used in nuclear reactors, spacecraft nosecones, and copper alloys.

The Canyon Mountains east of Delta, with their sharp summit peak, consist of long north-south strips of Precambrian and Cambrian sedimentary rocks, the Precambrian carried over the Cambrian on Sevier thrust belt faults. The west flank of the range is the breakaway zone for the Sevier Desert detachment, which lies deep underground to the west.

Climbing dunes almost bury Sand Mountain. Prevailing winds blow from the left in this photograph.

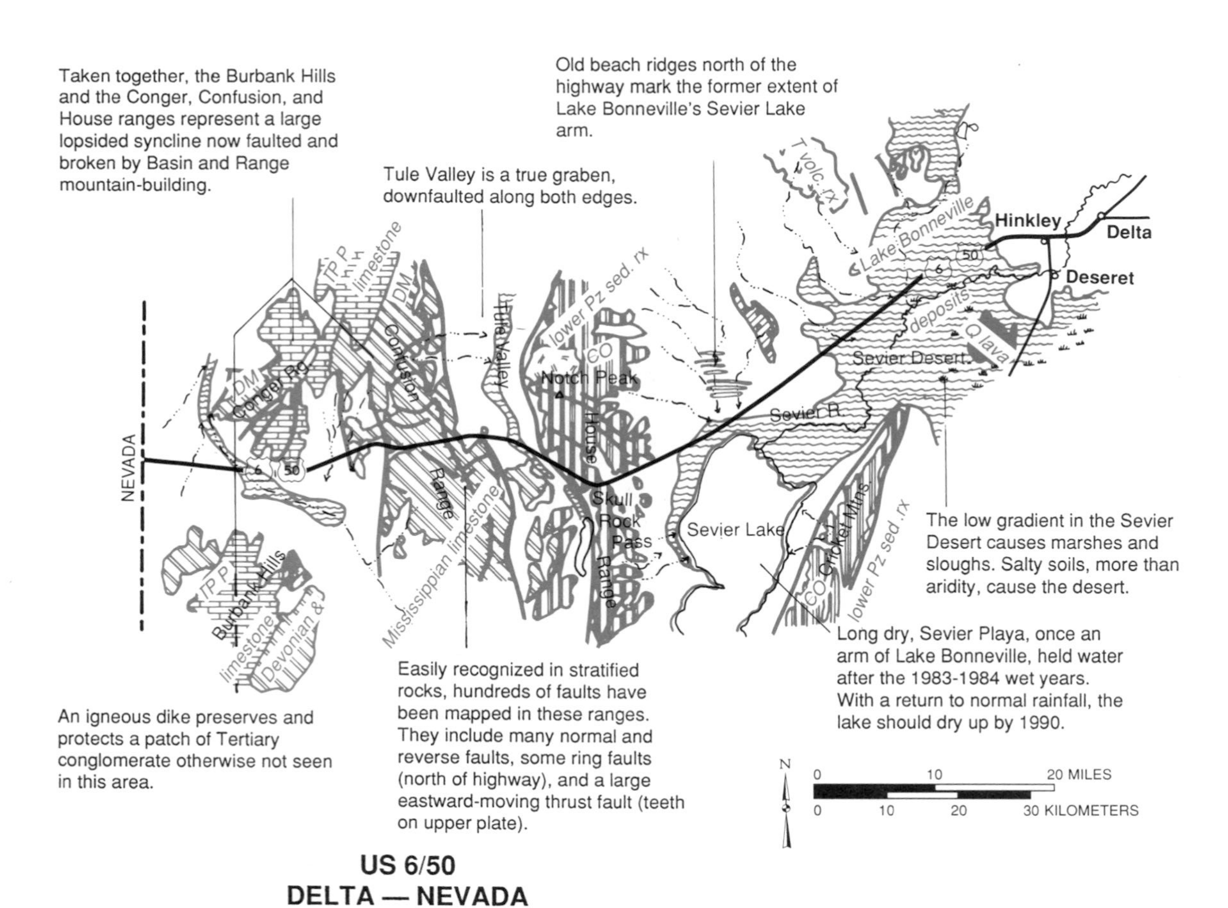

US 6/50
DELTA — NEVADA

US 6/50
Delta—Nevada
89 miles/143 km.

From Delta the highway runs southwest across the Sevier Desert, which is mostly surfaced with Lake Bonneville deposits — sandy delta sediments near Delta, finer lake-floor silt farther west. The bed of the Sevier River, which heads in the mountains of central Utah, twists across the level desert floor just south of the highway. The river rarely flows, partly because some of its water is diverted for irrigation, partly because remaining water sinks into the valley sediments or evaporates in the hot, dry climate.

In places, though, the desert surface is marshy — a seeming contradiction in terms! But the marshes are too salty to support much life. Elsewhere, low shrubbery tolerant to salt and able to survive the searing summer sun dots the barren surface; plants are more abundant near the highway, where they benefit from fresh water draining off the pavement.

The Drum Mountains northwest of Delta are edged on the east by a cluster of normal faults, many of which have left recognizable scarps on alluvial fans near the mountains. Fault scarps on such young surfaces indicate very recent movements. Seismic studies suggest that these nearly vertical faults may connect, about seven miles below the surface, with a detachment fault that extends almost horizontally below the Sevier Desert. Hills and mountains to the west, the House Range, bear the shoreline scars of Lake Bonneville, confirming that this valley, too, was part of the great Pleistocene lake. Comparison of elevations of these shorelines with their counterparts near Great Salt Lake shows that these are slightly lower than those that formed nearer the deepest part of Lake Bonneville — the part occupied by Great Salt Lake today. The weight of Lake Bonneville's thousand or more feet of water, much greater than that of today's Great Salt Lake, depressed the Earth's crust, especially near the deepest part of the lake. Since Lake Bonneville's demise, rebound of the crust slightly lifted the old shorelines.

From milepost 60, Sevier Lake, which appears on many road maps as Sevier Dry Lake, can be seen to the south. Until 1983, it was dry most of the time. During the wet years of the early 1980s, the Sevier River for the first time in living memory carried large amounts of water into Sevier Lake, turning the glaring white playa into a shimmering though shallow body of water. It may well be dry again by the time you read this.

Plans are afoot to mine chemical brines at the south end of the lake. Pumped from the subsoil, the brines will first be allowed to evaporate in the dry desert air, forming a flat salty floor for evaporating pans in which other minerals, also present in the lake brines, can be concentrated.

Cobbly gravel in quarries near the highway between mileposts 58 and 57 are stream deposits laid down along Lake Bonneville's old shore.

Hills to the north bear dark bands of tilted Paleozoic sedimentary rock—limestone, dolomite, and shale. In this area, north-south bands of Paleozoic rocks, older in the east, younger in the west, form successive ranges. The House Range, west of Sevier Lake and both north and south of the highway, contains Cambrian and Ordovician strata. The Confusion Range farther west consists of Silurian and Devonian strata. And at the northwest end of the Confusion Range and in the smaller Conger Range still farther west are Mississippian, Pennsylvanian, and Permian rocks. Well exposed and containing many fossils, the rocks in these ranges have been extensively studied.

Sevier Lake was a salty playa. In 1983 and '84 the Sevier River flowed into it, creating a shallow body of salty water that by 1985 began to dry up. Lake Bonneville beach ridges mark the foreground.

Notch Peak, at the crest of the House Range, consists of thick layers of Cambrian dolomite of the Notch Peak formation.

They document the history of this part of North America in Paleozoic time, when a western sea encroached upon the continent.

At milepost 41, the mountains ahead are the Confusion Range. To approach them our route crosses the southern end of Tule Valley, which has no drainage outlet. Water entering it sinks into the broad alluvial aprons around its margin or flows via numerous washes into a small playa . Lake Bonneville shorelines mark the mountain flanks, showing that this valley held an arm of the Pleistocene lake. The valley is a half graben, hinged down along a fault that edges its east side.

From Tule Valley, rounded pink rocky slopes come in sight to the north, just beyond Notch Peak. This part of the mountain looks different from that made of Paleozoic rocks; no stratification can be seen on it. It is a Jurassic intrusion that pushed up through Paleozoic strata.

The Confusion Range to the west, as well as small hills within Tule Valley, are composed of Silurian and Devonian sedimentary rocks. At about milepost 26 we cross a thrust fault that brought Silurian rocks east over Devonian rocks, reversing their normal youngest-over-oldest order.

Crossing the summit between mileposts 23 and 22, the highway descends into the Ferguson Desert, which drains west into the Snake Valley. Broad alluvial fans and aprons, so typical of Basin and Range country, surround the Conger Range to the northwest. Pennsylvanian and Permian rocks of this little range reach almost to the bend in the highway near milepost 19. They are the youngest of the Paleozoic rock units here.

The Snake Valley also lacks an outlet stream. Its playa, Salt Marsh Lake, lies about 30 miles north of this route. Shorelines marking the mountain flanks tell us that we are still within the domain of Lake Bonneville. This is its westernmost extent in this part of Utah. Farther north, it reached about 10 miles into Nevada.

West across the Nevada line is the Snake Range, now part of Great Basin National Park. The range is another metamorphic core complex, a mountain whose surface shows the crushed, broken, and re-crystallized rock of a major detachment fault. Relieved of the weight of rocks that once lay above it, the buoyant core of the mountains rose above its surroundings, bringing the detachment fault surface into view.

Utah 21
Nevada—Beaver
107 miles/172 km.

Wide alluvial aprons surround the Burbank Hills of western Utah and meet with similar aprons sloping from the Snake Range in Nevada. In places these aprons, partly pediment and partly gravel valley fill, are truncated at their lower ends; there you can see the gravel of which they are made.

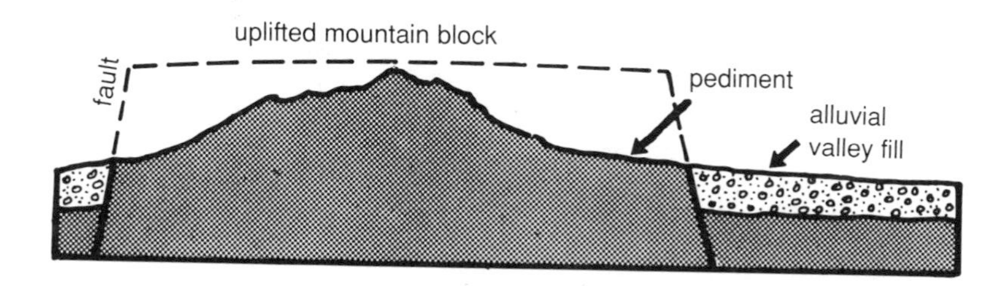

Pediments are eroded into the bedrock of mountain blocks. Alluvial aprons or bajadas are made of sand and gravel washed from the mountains and deposited between the ranges. Surfaces of pediments and bajadas are often continuous with each other, making it difficult to know where one ends and the other begins.

The deep valley between mileposts 3 and 4 in all probability formed in the last 100 years, in an episode of erosion usually blamed on the advent of cattle grazing and the resulting removal of protective plant cover.

Rocks exposed in the Burbank Hills and Mountain Home Range to the south contain Paleozoic sedimentary rocks arranged in a great syncline that extends south for 30 miles and north for more than 50 miles. As always with synclines, the youngest rocks are at the center. Devonian rocks appear near Garrison, Mississippian strata near Pruess Lake, and a wide band of Pennsylvanian and Permian limestone and shale in the Burbank Hills between mileposts 7 and 15. Light gray Pennsylvanian and Permian limestone comes quite close to the highway near mileposts 16 and 17, where the road bends slightly east between the Burbank Hills and the Mountain Home Range to the south.

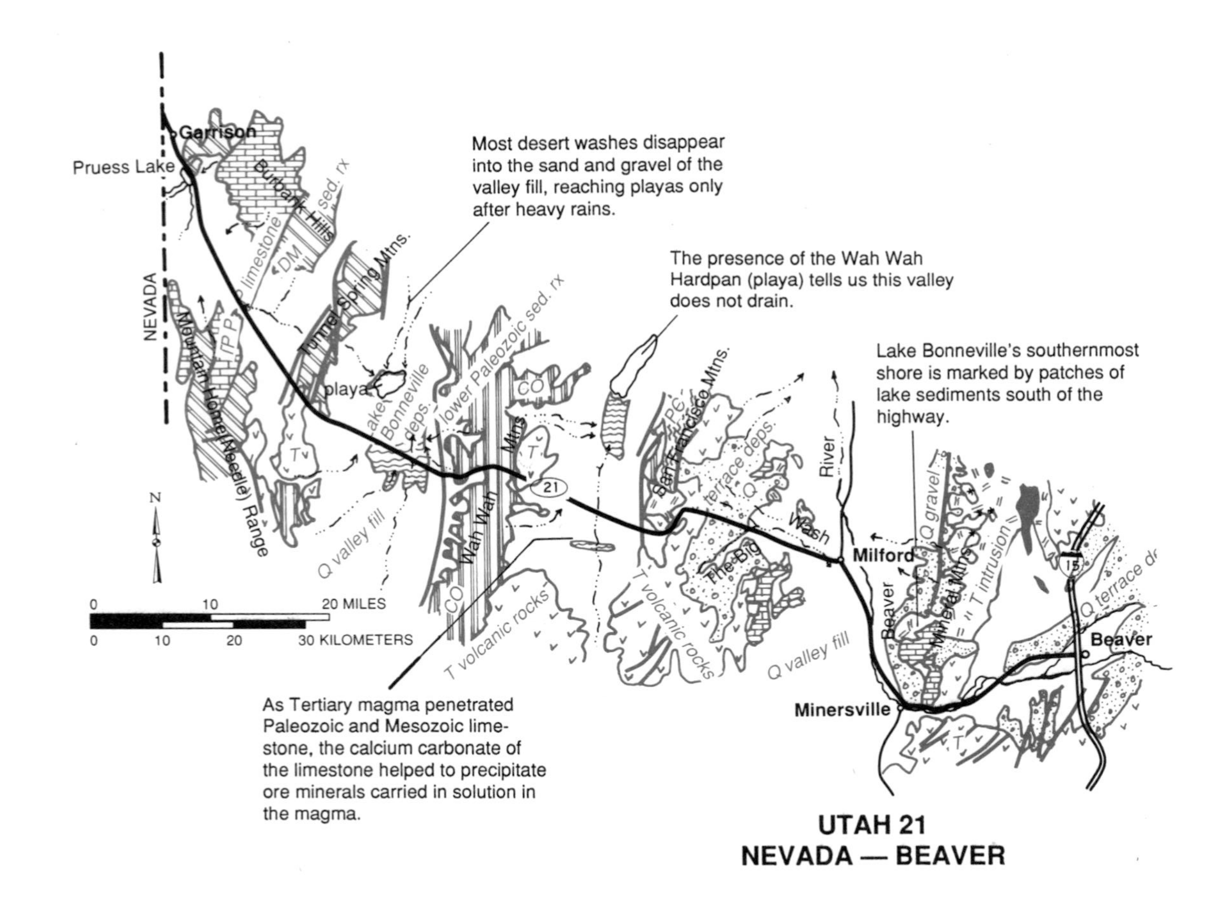

UTAH 21
NEVADA — BEAVER

Pruess Reservoir south of Garrison collects scanty runoff from surrounding hills. In the background is the Snake Range in Nevada, now part of Great Basin National Park.

We cross the northeast trend of the syncline at a slant, as can be seen on the map. Southeast of the pass we encounter older rocks again — first Mississippian and then Devonian; near the highway they lie beneath the gravels of the alluvial fan. Devonian rocks are exposed north of the highway at the next low pass, between mileposts 24 and 25. South of the highway are both Devonian sedimentary rocks and Tertiary volcanic rocks.

The long serrate ridge known as The Needles is Ordovician limestone, as are the rocks north of the highway and in highway cuts near milepost 26. Pale rock is Tertiary volcanic tuff.

Practically all the Paleozoic rocks here are marine limestone, thousands of feet of it, all deposited in the rapidly subsiding region west of the Paleozoic hingeline. Marine limestone is commonly made of infinite numbers of tiny animal shells, crushed, recrystallized, and for the most part unrecognizable. Larger, thicker shells retain their identity and can be found as fossils.

The highway descends into Pine Valley, a closed basin with its playa far north of the highway. In Pleistocene time this valley was part of Lake Bonneville, whose shorelines still mark surrounding mountains, especially the Wah Wah Mountains to the east. Lake Bonneville sediments cover parts of the valley floor.

The Wah Wah Mountains are almost entirely east-tilting Cambrian strata, with younger Paleozoic rocks beneath a thrust fault at the south end. Along their eastern side are some Tertiary volcanic rocks. Highway 21 crosses a low saddle in the middle of the range, giving us a good look at these strata. We encounter the oldest ones first, exposed low down on the western slope of the mountains: a thick layer of quartzite derived from almost pure quartz sand deposited along the beaches and bars of the Cambrian sea as it crept eastward across the subsiding continent. Above the quartzite, forming most of the western face of the range, are layers of Cambrian shale and limestone only slightly younger than the quartzite. Near and east of the saddle are still younger Cambrian rocks, predominantly limestone. Taken together, these three units record the gradual submergence of the west edge of the continent and gradual deepening of the Cambrian sea. Cambrian rocks are widespread in Utah and adjacent states, always represented by these three units: sandstone or quartzite at the base, then shale and shaly limestone, then pure marine limestone at the top.

East of milepost 45, the highway descends rapidly into the Wah Wah Valley, also once occupied by an arm of Lake Bonneville. Northward the valley is virtually continuous, except for a low sill, with the Sevier Desert, a much larger basin floored with Lake Bonneville deposits. Tertiary volcanic rocks, both lava flows and volcanic ash or tuff, close the southern end of the Wah Wah Valley.

The San Francisco Mountains east of the valley consist mostly of Precambrian schist. A Tertiary intrusion cuts through Cambrian and Ordovician sedimentary rocks at their south end. Precambrian rocks at the summit of the range have been pushed over Ordovician rocks along a thrust fault that is exposed low on its western slope. Recent research suggests that the rocks of this range once lay above the Mineral Mountains metamorphic core complex 25 miles to the east. If so, the Precambrian and early Paleozoic rocks moved 25 miles west on a detachment fault, part of the stretching out of western Utah.

Note the well developed alluvial aprons or bajadas that surround the mountains. The old mining camp of Frisco is near the south end of the range; mine workings can be seen from the highway. Mineral enrichment occurred when fluids from the intrusion reacted with Paleozoic limestone and dolomite — a common situation in Utah. Lead, copper, uranium, beryllium, and fluorite occur here.

East of the San Francisco Mountains the highway descends into Milford Valley, an intermountain basin unlike those farther west in that it drains north via the Beaver River, and so does not have a playa

Old lead and copper mines dot the southern part of the San Francisco Mountains. Mine dumps contain rock from the mines and overburden moved when gravels of an alluvial fan, still partly visible at right, were removed.

at its low point. In wet years, the Beaver River flows into the Sevier Desert and Sevier Lake.

Small washes from the surrounding San Francisco Mountains to the west and the Mineral and Tushar mountains to the east are also usually dry; when they flow their water sinks almost immediately into the porous valley fill. Among these tributaries is The Big Wash, just east of milepost 69. Gully walls display sandy and silty Pleistocene stream deposits.

Small ranges north and south of the highway between The Big Wash and Milford consist of bent and broken Paleozoic sedimentary rocks and several small Tertiary intrusions. Again we see mineralization where the intrusions came in contact with Paleozoic limestone.

We have seen no Jurassic or Cretaceous rocks along this route. They do occur in the southern Mineral and Wah Wah mountains. Mountain uplift in Cretaceous time raised this region well above its surroundings, creating the ranges known to geologists as the Sevier Mountains. There, erosion rather than deposition ruled.

Milford is an agricultural center. Water for its fields comes from Minersville Reservoir near the Mineral Mountains, as well as from wells. The water table is dropping because wells are producing water faster than it is replaced, a practice that geologists call mining the water.

The large central mass of the Mineral Mountains east of Milford is granite, a large Tertiary intrusion. An unusual line of little Quaternary volcanoes perches along the crest of the range. Contorted and faulted Paleozoic and Mesozoic sedimentary rocks at each end of the range are part of the upper plate above a detachment fault. About 10 million years ago the upper plate broke away and slid west off the large granite intrusion. Because of its lesser density, the granite mass, relieved of its burden, rose rapidly to form the metamorphic core complex range.

From Milford the highway follows the Beaver River upstream, circling south of Bradshaw Mountain. The Black Mountains south of Minersville are remnants of Tertiary volcanoes built up of sloping layers of lava, tuff, and breccia. These mountains extend west from a volcanic region covering thousands of square miles. Between mileposts 92 and 93, the change in dip (from west to east) of some of the lava and ash flows shows that the highway passes through the heart of one of the individual volcanoes.

The Tushar Range above the town of Beaver is part of the same volcanic region. It is a cluster of closely spaced composite or stratovolcanoes that erupted in Tertiary time. The volcanoes had violent histories, with huge explosive eruptions that far surpass any known in historical time. The volcanic rocks include quantities of breccia, volcanic ash, and lava flows. Some of the breccia originated in massive mudflows that streamed down the flanks of the volcanoes to inundate the surrounding country.

Near Beaver the highway crosses a terrace of Pleistocene stream sediments and then descends across successive shoreline ridges that resemble those of Lake Bonneville. They were formed by a smaller lake that occupied Beaver Basin in Pleistocene time.

IV
Something Special
National Parks and Monuments

Almost all of Utah's national parks and monuments are primarily geologic in nature. Many also contain evidence of human history in ruins and petroglyphs. All have interesting and varied plant and animal life. Thanks to high elevation and dry climate, the rocks stand out in bold relief.

Most of these geologic parks and monuments — Arches, Canyonlands, Capitol Reef, Natural Bridges, Rainbow Bridge, and Zion — lie in the Colorado Plateau region. Bryce and Cedar Breaks ornament the abrupt edges of the High Plateaus. Dinosaur National Monument lies in the no-man's land between the Colorado Plateau region and the Uinta Mountains. And Timpanogos Cave is in the mountains proper, near the southern end of the Wasatch Range. None of Utah's national parks are in the Basin and Range region, though across the state line in Nevada is Great Basin National Park, one of the newest members of the National Park System.

Many park areas have interesting roads and trails. Some have nature trails with guide leaflets, as well as scheduled talks, introductory movies or slideshows, and nature walks led by park personnel.

Please do not collect rocks or fossils in the national parks and monuments. They are protected there for your enjoyment; cooperate in preserving them for others to enjoy. And do not deface the rocks that Nature has so beautifully exposed. Fossil hunting is also prohibited on all other public land — including Bureau of Land Management and National Forest areas.

Arches National Park

More than 200 stone arches, all of them in Jurassic dune-deposited sandstone of the Entrada formation, are the main attractions of this park. Many are visible from park roads, others from trails. Most are in narrow fins of coral-colored rock sliced by numerous parallel joints. And most owe their origin to two unusual anticlines cored with salt and gypsum.

Unlike most rocks, gypsum and salt can actually flow, very slowly and very gradually, moving much as ice does in glaciers or as red-hot iron does under a blacksmith's hammer. In addition, salt and gypsum are less dense and therefore more buoyant than other rocks. Like oil in water, they do their best to rise toward the surface.

Thick layers of salt and gypsum, interlayered with sedimentary rocks, were deposited in this region in Pennsylvanian time, when sea water evaporated in a shallow, almost landlocked bay. Subsequent layers of sandstone, siltstone, mudstone, and limestone soon covered the salt and gypsum, and at first held them in place despite their tendency to rise toward the surface. But overlying rocks were a little thinner, and pressures therefore slightly less intense, over a number of northwest-southeast trending ridges caused by faulting in Precambrian rocks. Above these ridges the salt and gypsum succeeded in pushing upward quite strongly, to dome overlying sedimentary rocks into anticlines. Because thinning of overlying sedimentary layers was controlled by the position of the old fault ridges, today's salt anticlines trend in a north-west-southeast direction. At least a dozen large salt anticlines exist in eastern Utah and western Colorado.

The narrow fins in which so many arches appear were created as the salt-cored anticlines formed, when rocks overlying the salt and gypsum broke along northwest-southeast joints. The joints, and therefore the fins, are best developed in the Fiery Furnace and Devils Garden areas, where most of the arches occur.

As rain and snowmelt dissolve minerals that hold sand grains together, joints between these rocks widen, a process helped by plant roots, soil moisture, and freezing and thawing. Where some parts of the fins are weaker and weather more rapidly, alcoves and arching caves form. If weak zones extend clear through a fin, two deepening caves on opposite sides of the same fin may join, creating a window or an arch. The opening is then enlarged by gradual flaking or spalling of rock material as it slowly weathers, or by sudden falling of unsupported rock.

The terms arch, window, and natural bridge are often misused. A window, like a window in a house, is a small opening usually well above ground level. An arch is larger, and spans a rock surface; it forms with little or no help from streams. The term natural bridge should be used only for spans created by stream erosion, with watercourses (even dry watercourses) running under them.

In early Jurassic time eastern Utah was part of a great desert comparable in many ways to today's Sahara. The continent was farther south then, at latitudes where most of the world's great deserts are today. On the Jurassic desert were deposited the dunes that eventually became the thick, light gray, cross-bedded Navajo sandstone, which forms humpy bare-rock surfaces in parts of the park today, particularly near the entrance road and in the Windows area.

After the Navajo sandstone had been deposited, rivers for a time brought in finer, softer, floodplain types of sediments — silt and clay that now make up the lower part of the Entrada formation, the reddish, crumpled and distorted layers below today's arches. Then, desert conditions returned, and more dunes marched across the river floodplains, just as Sahara sand dunes at times march out onto the Nile floodplain today.

Arches' arches are in coral-colored dune sandstone, the upper part of the Entrada formation, that signals a return to desert conditions. In places the weight of this sand distorted the still-soft mudstones of the formation's lower part; you can see the distortion at the base of many of the arches in the Windows section of the park.

Even as these rocks were being deposited, layers of underground salt and gypsum were sporadically pushing up along the lines of ancient faults, creating a series of narrow anticlines across the Utah-Colorado border. Later, as the Rocky Mountains rose in Colorado, Wyoming, and central Utah in late Cretaceous and early Tertiary time, thick blankets of

Like other arches in the national park, Delicate Arch is in the upper part of the Entrada sandstone. Swirls of wind-deposited sand festoon surrounding rocks.
—Ray Strauss photo.

sand, silt, and clay, much of it deposited in large inland lakes, covered the dune-formed sandstones and associated river deposits. These blankets have now eroded off the immediate area of the park, but can be seen to the north in the Book Cliffs and Roan Cliffs.

As regional uplift occurred, groundwater dissolved the salt and some of the gypsum of the salt and gypsum anticlines, flushing them down the Colorado River, which flows along the south edge of the national park. Crests of the anticlines collapsed, dropping several hundred feet to create fault-edged valleys called grabens. Within this park, the two collapsed anticlines are now Cache Valley Graben and Salt Valley, both accessible by road. The valleys are bordered by rocks that dip away from them in both directions — the two flanks of the old anticlines. Younger rocks, irregular masses of Cretaceous shale and sandstone, collapsed into them as the salt eroded away. In places the valleys still contain large masses of contorted gray gypsum.

Many interesting rock details are exposed in this park. Bright green rock near the Wolfe Cabin in Cache Valley Graben is colored by an unusual iron mineral. Springs not far from the cabin furnished the Wolfe family with palatable drinking water, much less salty than that in the stream near the cabin.

The trail to Delicate Arch crosses smooth, barren sandstone slopes marked with little solution furrows and potholes that hold rainwater for a time, and that harbor tiny animals and plants able to live out their entire life cycles in the few days or weeks after a rain. Acids that these organisms excrete speed up solution of the calcium carbonate that cements the sandstone, helping the potholes to enlarge. The La Sal Mountains, a cluster of Tertiary intrusions, tower to the south, their summits frequently patched with snow.

Courthouse Towers and adjacent cliffs give good exposures of the lower, crumpled, easily eroded mudstone at the bottom of the Entrada formation, and the massive, resistant dune sandstone above it. In places the massive sandstone has broken along vertical joint planes; many of the flat surfaces are darkened with desert varnish. The head of the Egyptian Queen, along the Park Avenue Trail, is thought to have shifted during an earthquake.

Bryce Canyon National Park, Cedar Breaks National Monument

In both of these parks, huge natural amphitheaters are carved into soft sedimentary rocks and decorated with towers, turrets, and crenelated ridges so intricate that they must be seen to be believed. Soft pink and white siltstone and limestone layers that wall the amphitheaters, formerly assigned to the Wasatch formation, are now recognized as the Claron formation, deposited in a Paleocene lake. Lifted high on two of Utah's high plateaus, these rocks have been, and still are, exposed to headward erosion by branching streams. They are also subject to freeze-and-thaw weathering, the pelting force of rain and hail, landsliding, and wind. Studies of the growth rings of trees along the canyon rims, especially those whose roots now hang out over the rims, show that canyon rims are receding 9-48 inches per century, geologically quite a rapid rate.

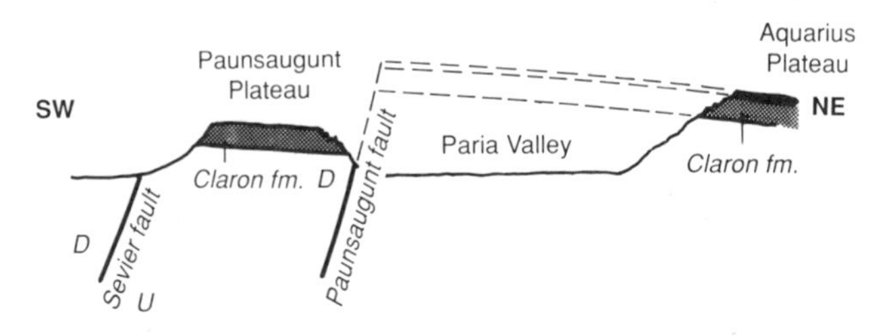

Wide Paria Valley was carved in the uplifted block of the Aquarius Plateau. Eroded Pink Cliffs of Bryce are repeated on the edge of this higher plateau.

The Bryce Canyon amphitheater, formed by the many-branched headwaters of Fairyland, Campbell, Bryce, and Cathedral creeks, developed on the east side of the Paunsaugunt Plateau, fault-edged on east and west. The rim of Bryce's great amphitheater is on the dropped side of the Paunsaugunt fault.

Cedar Breaks developed on the west side of the Markagunt Plateau, where it drops off abruptly along the Hurricane fault, which, with several thousand feet of movement, separates Utah's high country from the Great Basin region to the west. In this case the dropped side of the fault is the low side topographically.

Many lesser faults parallel or cut across the main faults at both sites. And many joints are present, which with the faults have helped to govern patterns of erosion.

Water, rain and snow and runoff, is of course the primary agent of erosion here. Streams spring to life after heavy summer thundershowers or during winter snow-melt, to cut downward and headward into the soft rocks. In this region, particularly at high altitudes near the rims of the amphitheaters, day-to-night temperature changes bring about repeated freezing and thawing of water that has seeped into joints and pore spaces in the rock, crumbling the poorly cemented lake sediments. At these elevations, overnight freezing and daytime thawing occur as often as 200 to 300 times a year.

As you can tell by looking at the walls of Bryce Canyon or Cedar Breaks, horizontal sandstone, limestone, and siltstone layers of the Claron formation vary in resistance to erosion. Siltstone layers are the most easily eroded. Sandstone, limestone, and dolomite (a rock similar to limestone except

Cedar Breaks, seen here from Spectra Point, is carved by headward erosion of a relatively small stream. Oligocene lava flows, quite a bit younger than the lake deposits, cover Brian Head at top center. —Ray Strauss photo.

containing more magnesium) are more resistant, and form the canyon rims and prominent ledges seen in both Bryce and Cedar Breaks.

The intermountain lake in which these rocks were deposited measured about 250 miles long and nearly 75 miles wide, a little larger than Lake Erie. After the sediments had accumulated, around 55-60 million years ago, uplift of the area surrounding and including the Plateau country raised these rocks to about their present elevation. Since then, erosion along tributaries of the developing Colorado River created the landscape we now know.

A closer look at the Bryce amphitheater and the much larger Paria amphitheater of which it is a part, shows gently sloping mesas far out on the floor of the Paria amphitheater, beyond the town of Tropic. These mesas are remnants of a once higher amphitheater floor. After a period of stability, perhaps as recently as 500,000 years ago, the Paria River and its tributaries cut downward into this floor, lowering the level to which erosion of the Paria amphitheater could proceed. Bryce Creek

Rocks of the Claron formation, deposited in a Paleocene lake, have eroded into the colorful towers and turrets seen at Bryce Canyon National Park.
—Ray Strauss photo.

cannot cut below the level of its confluence with the Paria River, near the town of Tropic; the Paria River in turn cannot cut below the level at which it joins the Colorado River below Glen Canyon Dam.

Viewpoints along the rim drive introduce visitors to varied aspects of both Bryce and Cedar Breaks. Hiking along one or more of Bryce Canyon's many trails is a good way to become even better acquainted with the ornately carved ridges and turrets.

At Cedar Breaks, the Claron formation is thicker and quite a bit more colorful than at Bryce, and is often called the Cedar Breaks formation. Brian Head, just southwest of Cedar Breaks, is capped with Oligocene lava flows and volcanic ash. A bonus to Cedar Breaks visitors comes in the form of much younger lava flows, some of them looking as if they erupted only yesterday, along Utah 14 east of the national monument.

Canyonlands National Park

The layer cake geology that characterizes the Plateau region nowhere shows up better than in Canyonlands National Park. All of the rocks are sedimentary; most are flat-lying, with alternating hard and soft layers. Alternating cliffs and slopes formed by these layers are fully exposed on many mesas, buttes, pinnacles, and canyon walls. The brilliant hues of many of the rock layers, pinks and brick reds and salmons, are due to varying amounts of iron oxide minerals.

Though they look essentially horizontal, the layered rocks here actually arch across Canyonlands in a broad, gentle bend, the north end of the Monument Upwarp of the Arizona-Utah border. Sedimentary rocks exposed within the national park were deposited during Pennsylvanian and Permian time and during Triassic and Jurassic time. The oldest strata are exposed nearest to the Colorado and Green rivers.

In the Needles area, sandstone pinnacles worn by rain and wind dwarf a pair of hikers. —Ray Strauss photo.

Near the Colorado River, the strata bow up more sharply, probably as a result of downward erosion by the river and the upward push of salt and gypsum from thick Pennsylvanian sediments below the surface. Salt and gypsum, which become plastic under pressure, are less dense and therefore more buoyant than other rocks; they constantly seek to move up against whatever rock overlies them. Normally, the strength of overlying rocks suffices to keep them in place, but here where the river has removed some of the overburden, the salt and gypsum have flowed up, gradually arching the overlying sediments. Additional salt and gypsum flowed sideways toward the river, causing some degree of collapse where they had been, creating the Grabens, parallel-walled fault valleys in the southern part of this park.

Where did the salt and gypsum come from? These minerals, along with potash and anhydrite, are known as evaporite minerals because they form where sea or salty lake water evaporates. In Pennsylvanian time, this region was a narrow arm of the sea, an almost landlocked bay, with restricted circulation that allowed for evaporation of sea water yet permitted its constant replenishment. Layer after layer of evaporites, a total of 3000 feet, formed here during Pennsylvanian time. Then the layers were covered with the river muds and sandstones now visible in the walls of Cataract Canyon near and below the confluence of the Green and Colorado rivers. Beach and dune and alluvial fan deposits were laid down during Permian time: the striped pink and white Cedar Mesa sandstone visible in the Needles area, the Cutler formation shales and siltstones of other parts of the park, and the White Rim sandstone.

Triassic rivers and streams delivered more burdens of mud and sand to the area, thick floodplain and delta deposits that now make up the Moenkopi and Chinle formations. These soft, poorly consolidated deposits are notable slope-formers, and have eroded back from benches of Cedar Mesa and White Rim sandstones. In doing so, they undermined the Wingate sandstone, the massive Jurassic rock that forms the high walls of Island in the Sky and the sheer cliffs around the margins of the park.

Triassic and Jurassic rocks are well exposed along the southeast park entrance. Here, a tall red cliff of Jurassic Wingate sandstone rises above slopes of Triassic shale. Above the Wingate are red floodplain and delta deposits, the Kayenta formation. White rock at top is Navajo sandstone.

The Wingate sandstone introduces us to another aspect of the region's history. In a desert region the size of today's Sahara, sand dunes accumulated in early Jurassic time to become the Wingate and Navajo sandstones. The Kayenta formation, a ledgy band of siltstone and shale separating these two formations, shows us that river deposits formed during an interlude between the two sand dune advances, as perhaps today's Nile River might, if it changed course, cause an interlude in dune deposits of the Sahara Desert.

As seen from Grand View Point and Island in the Sky, the many bends of the Green and Colorado rivers are entrenched in meanders inherited from Miocene time, when the rivers looped lazily across a gently sloping plain. From these vantage points one can see the La Sal, Abajo, and Henry mountains, each range a cluster of igneous intrusions that domed up layered rocks now eroded from their crests. The Shafer Trail, built for access to uranium mines, descends one side of Island in the Sky, going through almost all the rock units exposed in the park.

At Newspaper Rock near the southeast entrance to the park, figures were pecked through dark desert varnish to expose lighter sandstone beneath.

Upheaval Dome, northwest of Island in the Sky, is an unusual mound of Triassic and Jurassic sandstone and mudstone surrounded by concentric rings of younger sedimentary rock. Long credited to upward flow of salt and gypsum, it is now thought to be the result of a large meteorite impact, probably in late Cretaceous or very early Tertiary time. The shock wave generated by the impact is thought to have created a short-lived underground cavity into which surrounding rock layers, highly fractured by the impact, then slid, converging and surging upward at the cavity's center. At the time it formed, the present structure lay one to two miles below the surface; overlying rock has now been eroded away.

In the Needles District, 4-wheel-drive roads penetrate narrow canyons and lead to several stone arches. One of them leads to the Grabens and the confluence of the Green and Colorado rivers.

Chesler Park in the Needles District is surrounded by turrets and fins carved in red-and-white-striped sandstone. Erosion has benefitted from the many vertical joints. —National Park Service photo by M.W. Williams

The Navajo, Wingate, and Kayenta sandstones weather into rounded bluffs, ledges, and domes along the summit of Capitol Reef. Holes in the rock at lower left are tafoni, thought to have formed by wind erosion. —Ray Strauss photo.

Capitol Reef National Park

Waterpocket Fold, a single large monocline formed in late Cretaceous time, is the dominant feature of this park. Prominent parts of the fold, the Capitol Reef itself, consist of steeply tilted Jurassic rocks: the dune-formed Wingate sandstone, thin-bedded floodplain deposits of the Kayenta formation, and the massive dune-formed Navajo sandstone, which forms the crest of Capitol Reef.

Triassic and Permian strata appear west of the reef, and younger, Cretaceous strata are well exposed east of it. Though these rocks dip less steeply, they are also to some extent involved in the monocline that forms the reef.

Since sandstone, siltstone, mudstone, and limestone layers each respond differently to erosion, the fold appears as a series of ridges, some quite high, some relatively insignificant. The most prominent ridges are those formed by the Wingate and Navajo sandstones — the former making up the angular cliff on the west side of the reef, and the latter forming both the rounded white summit domes that give the reef its name, and its bulbous eastern slope. Waterpocket Fold is about 100 miles long in a north-south direction, but less than three miles wide.

As seen from the west, the cliff of Wingate sandstone, streaked with lichens and darkly splotched with blue-black desert varnish, appears at the top of the rock sequence. Breaking along vertical joints, it maintains its sharp profile while contributing to talus slopes below. The talus rests on a slope of Triassic rock — the many-hued Chinle formation, rich in clays derived from volcanic ash. The bottom of the slope is the dark brick-red mudstone and siltstone of the Moenkopi formation, which also surfaces most of the park west of Capitol Reef. In places the Moenkopi formation contains veins and veinlets of selenite, a lustrous, silky form of gypsum. Watch for it along the scenic drive south of the visitor center.

Exposed in Sulphur Creek Canyon and other deep gorges west of the reef are light gray ledges of marine limestone that correlate with the Permian Kaibab limestone on the rim of Grand Canyon. Below them is a little cross-bedded sandstone of the Cutler formation.

Black lava boulders on the hill near the visitor center came from high lava-capped plateaus to the north and west; they were brought here by streams draining ice age glaciers. Grand Wash and the Fremont River carried some of these black boulders right through Capitol Reef.

Driving east through Capitol Reef along the Fremont River, or weather permitting, walking east through one of the other gorges that penetrate the reef, you'll see the Wingate sandstone first, and then younger rocks: ledgy, river-deposited sandstone and siltstone of the Kayenta formation, and massive white or pinkish, cross-bedded Navajo sandstone, a rock that weathers to rounded summits and slopes rather than to sharp-edged cliffs. This is the rock unit that forms the summit and eastern

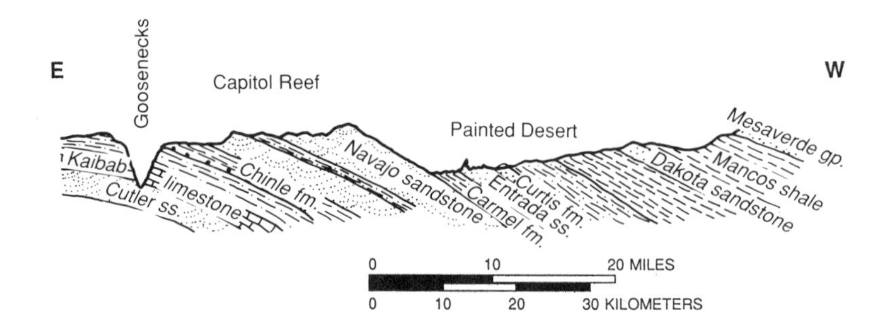

Section along Utah 24 through Capitol Reef

slope of Capitol Reef and that contains numerous natural pocket-like cavities that fill with rainwater, giving Waterpocket Fold its name.

Both the Wingate and Navajo sandstones display many clues to their dune origin. Both are cross-bedded, with long sloping laminae peculiar to dune sandstone. Both contain fine, rounded, even-sized, frosted sand grains. In both, wind-formed ripplemarks run up and down former dune slopes. (Water-formed ripplemarks always run at right angles to the slope on which they form.)

East of Capitol Reef are more Jurassic rocks: the deep red Entrada sandstone, eroding into unusual pinnacles and spires, and the valley-forming Morrison formation. Both the Triassic Chinle formation west of the reef and the Jurassic Morrison formation east of it contain abundant volcanic ash, which decomposes to bentonite, a swelling-when-wet and shrinking-when-dry clay that discourages plant growth. Both these formations also contain non-oxidized iron minerals that give them their purplish and greenish hues.

The Morrison formation contains pockets of uranium ore; small prospect pits dating back to the 1950s and '60s dot its line of outcrop east of Capitol Reef.

East of the Morrison formation valley is a stockadelike ridge of Dakota sandstone, the oldest Cretaceous unit here. Younger Cretaceous rocks surface the area farther east, dark gray valley-forming Mancos shale, and cliff-forming Mesaverde sandstone, interlayered with shale and coal.

Waterpockets in the Navajo sandstone form along joints that serve as natural channelways for running water. Most of them begin their development behind wind-etched depressions due to cross-bedding, where tiny, shallow pools collect after rainstorms. Standing water, even when it doesn't stand for long, weakens the rock just a little; wind or more running water removes loosened sand grains, slightly deepening the pools. As they increase in depth, tiny, short-lived plants and animals come to inhabit them, secreting acids that further weaken the rock. As the waterpockets deepen, grain by grain, they become large enough to hold water for a considerable length of time.

Honeycomb weathering, development of fist-sized holes called tafoni, is common in the Wingate and Navajo sandstones. This type of weathering may be initiated by tiny cavities resulting from solution of rock material; later they may be enlarged by the wind.

Massive uplands of Jurassic Navajo sandstone rise above ledges of Kayenta sandstone and massive cliffs of Wingate sandstone. —Ray Strauss photo.

Streaked with black lichens, the Navajo sandstone forms many of the cliffs in the national park. Horizontally bedded siltstone layers at the center of this photograph are the Kayenta formation. Boulders at bottom include dark gray lava from lava plateaus to the west. —Ray Strauss photo.

Capitol Reef contains a maze of narrow, tortuous canyons. Trails penetrate some of them, and lead also to a number of arches, natural bridges, and summit domes. If you can arrange a pickup on the other side, walk through Capitol Reef at Grand Wash or Capitol Gorge, looking at the rocks along the way.

Dinosaur National Monument

This national monument straddles the border between Colorado and Utah, as well as the northern edge of the Colorado Plateau, where flat-lying rocks of the Plateau country tilt up along the south flank of the Uinta Mountains. Tilted sedimentary rocks, Permian to Cretaceous in age, form a succession of flatirons, hogback ridges, and cuestas along the mountain front, each ridge separated from the next by a valley in one of the more easily eroded layers. Paleozoic and

Precambrian sedimentary rocks form the central mass of the Uintas, an east-west range that is essentially a single anticline, faulted along both margins as well as across and along its center.

Two major rivers come together within the bounds of this great anticline: the Green River, with headwaters in Wyoming, and the Yampa, with headwaters in Colorado. Joining, they twist through the southern ridges at Split Mountain, emerging near the famous dinosaur quarry that gives this monument its name.

The oldest rocks within the national monument are exposed in the Canyon of Lodore, where the Green River cuts down into partly metamorphosed Precambrian sedimentary rocks — quartzite, slaty shale, and marble of the Uinta Mountain group. Paleozoic sedimentary rocks, most of them marine, lie on top of these ancient strata. One non-marine formation, the thick, cross-bedded Pennsylvanian Weber sandstone, adds in no small way to the scenic beauty of this monument. Deposited as dunes along the shore of the Pennsylvanian sea, this buff-colored rock forms barren surfaces of parts of the Uinta Mountain anticline, and rises as spectacular, sometimes overhanging cliffs above deeply entrenched meanders of the Yampa and Green rivers.

Still younger rocks show us that early in Triassic time, when the land was low and nearly flat, much nearer to the equator than it is now, sediments accumulated in river floodplains and deltas, with the sea not far to the west. By early Jurassic time, when mountains to the west intercepted the rain clouds, the area had become part of a great desert that stretched from western Wyoming to the southern tip of Nevada.

Then, late in Jurassic time, the desert became a swampy coastal floodplain marked with shifting river channels and shallow lakes. Sandstone and mudstone deposited there now appear as the rainbow-hued Morrison formation. Dinosaur remains have been found in this formation at many localities in Utah and adjacent states. The dinosaur site within this national monument was discovered in 1909, and has been worked sporadically ever since, though bones are no longer removed from the rock in which they are entombed.

In the quarry face, dinosaur bones jut from the sandstone of a former river channel. More than 350 tons of bone have been removed from this quarry. Now, however, bones are left in place as they are uncovered.

The dinosaur bones were fossilized by gradual addition of silica from groundwater, so that now they are much harder than normal bone. The rocks in which they are enclosed tip upward steeply at the edge of the Uinta Mountains. The bones include the remains of at least 12 species of dinosaurs. Fossil crocodiles, turtles, frogs, and freshwater clams indicate that dinosaurs were not the only denizens of the Jurassic swamps.

Scenically, Dinosaur National Monument has much to offer besides the dinosaur quarry. The road to Harpers Corner, passing through Cretaceous, Jurassic, and Triassic rocks upturned along the mountain front, climbs the huge Uinta anticline, here surfaced with Weber sandstone. Much of the route is through barren slickrock terrain, where the cross-bedded dune deposits and flat, reddish interbeds of the Weber sandstone are prominently displayed. The road crosses the Yampa fault near the monument boundary. North of the fault several viewpoints overlook the Yampa graben, where the crest of the mountain has dropped some 2000 feet between the Yampa fault and the Mitten fault to the north. Harpers Corner also offers good views to the west, to Whirlpool Canyon, Island Park, and Split Mountain.

Hogbacks and cuestas along the southern edge of the Uinta Mountains, the eroded edges of Paleozoic and Mesozoic formations that once arched across the uplift, can be seen from the road to the Split Mountain campground.

In various places within the park, patches of soft, easily eroded gravel, sand, and silt show us that in early Tertiary time the entire area was deeply covered with sediments washed off the Rocky Mountains to the east and north — so deeply, in fact, that the Uinta Mountains were almost completely buried. At that time, both the Yampa and the Green rivers meandered across the surface of these sediments. With uplift of the entire region around five million years ago, the rivers cut down through the soft sediments, retaining their meandering pattern. By the time they reached the buried mountains, they were prisoners in their own canyons, and continued to carve downward, keeping the same meander loops, to form today's spectacular Yampa Canyon, Canyon of Lodore, and Whirlpool and Split Mountain canyons.

Natural Bridges National Monument

In this national monument, entrenched meanders of White and Armstrong creeks cut deeply into the Cedar Mesa sandstone, a pale, cross-bedded Permian sandstone particularly conducive to arch development. The large-scale dune cross-bedding in this formation is interrupted by weak horizontal layers of silt and clay deposited in low, flat areas between the dunes. Easily eroded, these interdune deposits help to undermine the stone

Most natural bridges develop where deeply entrenched streams undercut and break through the fins of rock that separate their meander loops. Once a passage has been opened, creating a natural bridge, the stream takes the more direct route.

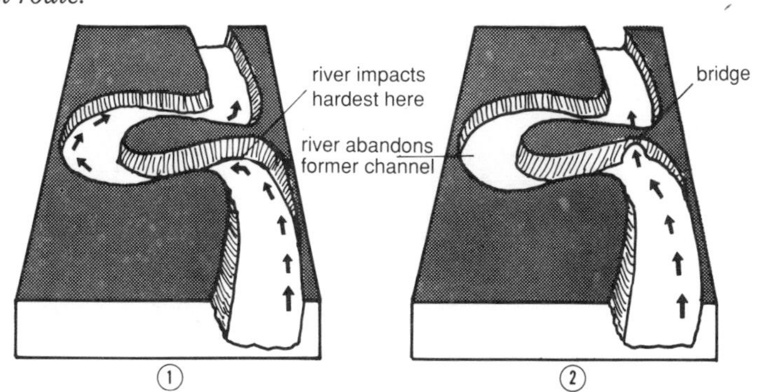

fins, creating the undercuts and alcoves that ultimately break through the rock fins.

Streams here, often dry, occasionally rush through their narrow canyons with considerable force, pounding the rock fins with sand, pebbles, and boulders — the tools of erosion. Once a passage has opened through a fin, the stream quickly adopts and enlarges the new channel. Rain and runoff from the plateau surfaces help with the job, as do repeated freezing and thawing, which loosen scabby flakes of rock as well as individual sand grains.

The three natural bridges in this national monument illustrate successive stages in bridge development. Kachina Bridge represents an early stage: thick and massive, with a relatively small passage below it. Sipapu Bridge is thinner, and the stream below it is no longer wearing away its abutments; further thinning will be at the hands of rain, snow, and freeze-and-thaw weathering. Owachoma Bridge, formed near the junction of Armstrong and Tuwa canyons rather than at a

Natural bridges, at first thick and massive, with time become thinner and thinner, finally becoming so delicate that they collapse.

meander loop, is in a late stage of development: slender and increasingly fragile. As these bridges continue to erode, and ultimately weaken and fall, others will form at sites where streams still pound the sides of rock fins.

Several other interesting geologic features exist in this national monument. The drainage pattern is controlled by two sets of joints: a northeast-southwest set, dictating the direction of White Canyon and many tributaries, and a northwest-southeast set, dictating the direction of Armstrong Canyon and its tributaries. In places, bare rock surfaces contain small potholes which, after rain, hold small pools of water. The little pools are well populated with tiny plants and animals, who help to enlarge their pools as the acids they secrete dissolve the calcium carbonate that cements sand grains together.

Tapered black bands of lichens mark seepage lines on many cliffs. Shinier blue-black surfaces are desert varnish.

The broad, sweeping cross-bedding visible in the Cedar Mesa sandstone betrays its origin in sand swept up gentle windward slopes of dunes and dropped on their steeper leeward slopes. In places, crumpled bedding interrupts the smooth sweep of the cross-beds, showing where sand slumped and slid down the steep lee slopes. The sandstone is so poorly cemented that you can rub grains loose with your hand. Even falling raindrops can erode such loose sand.

Rocks younger than the Cedar Mesa sandstone appear nearby. From bottom to top, the order in which they were deposited, they are: dark red siltstone and mudstone of the Triassic Moenkopi formation; the white ledge, often stained dark, of the Triassic Shinarump conglomerate, which with overlying purple and greenish gray shales makes up the Chinle formation; and capping the cliff to the north, the Jurassic Wingate sandstone.

Though it has been eroded off Cedar Mesa around White and Armstrong canyons, the Moenkopi formation contributes in a major way to the soft, puffy soils that surface the mesa, adding enough hematite to make it pink and enough gypsum to make it puffy.

The graceful arch of Rainbow Bridge, for many years accessible only by a long and arduous trail, can now be reached easily from Lake Powell.
—Ray Strauss photo

Rainbow Bridge National Monument

Spanning Bridge Canyon, a narrow gorge descending from the great dome of Navajo Mountain, the graceful 300-foot-high arch of Rainbow Bridge is considered the largest known natural bridge. In a remote wilderness of narrow canyons, precipitous cliffs, and bare-rock ridges, Rainbow Bridge developed in the Navajo sandstone, Utah's master scenery-maker. The long, sweeping cross-bedding in this formation, its fine, even-sized, rounded sand grains, and its ripplemarks like those on modern dunes, show it to be a sand-dune deposit.

At the time the Navajo sandstone was deposited, North America lay south of the equator, at the same latitude at which several of the world's great deserts — dry regions of prevailing easterly winds — exist today. Studies of the average dip directions of Navajo sandstone cross-bedding indicate that it, too, was deposited by easterly winds, but counterclockwise rotation of North America has shifted the dip direction to south.

Its long, sweeping cross-bedding, along with its uniformly fine-grained texture and the strength of the calcium carbonate that cements the sand grains together, make the Navajo sandstone particularly conducive to arch formation. Where it is undermined it tends to break away in long curves, many of which match the curvature of Rainbow Bridge. Erosion of soft layers of fine siltstone and claystone within the formation may initiate arch development.

At Rainbow Bridge, the Kayenta formation, the slabby, ledgy rock layer below the Navajo sandstone, was responsible for arch formation. The Kayenta formation is less porous and less permeable than the Navajo sandstone. A line of springs and seeps along the contact between the two formations shows that it doesn't absorb the rain water and snow melt that percolate down through the Navajo sandstone.

Behind the east buttress of Navajo Bridge is evidence of another factor in bridge formation: an abandoned gooseneck meander of an older Bridge Creek, a deeply entrenched meander inherited from a gentler stream course. Rushing toward this meander, the stream at one time threw its full arsenal of sand,

Where undermined, the Navajo sandstone falls away, leaving incipient arches on rock faces. Here near Lake Powell, thinly layered siltstone of the Carmel formation caps the thick sandstone layer.

pebbles, and boulders against the rock fin that jutted out into the meander bend, eroding and thinning it. Finally, the stream broke through the fin, whereupon it abandoned its old channel in favor of the shorter route toward the Colorado River.

At first Rainbow Bridge must have been massive and not particularly graceful, certainly not rainbow-shaped. A smallish opening probably guided the stream beneath a heavy, flat-topped span. But the Navajo sandstone tends to fall away in arches, and the Kayenta formation is relatively impermeable. Springs at the top of the Kayenta formation kept the porous sandstone wet and subject to freeze-and-thaw weathering, while also washing some of the cementing material from the sandstone. So gradually the opening enlarged.

As slabs of sandstone were loosened by weathering and undermining, thin, curving sheets and probably some large blocks of Navajo sandstone fell away, giving the natural bridge its arching contour. At some stage the stream below it cut down into the Kayenta formation, effectively heightening the bridge. Weathering continues to perfect the rainbow span, and will very gradually narrow it until it will no longer be able to support its own weight.

Part of the scene at Rainbow Bridge is the nearby dome of Navajo Mountain. The mountain is a laccolith, an intrusion of igneous rock that domed up sedimentary layers. Largely eroded from its summit, these layers still encircle its flanks; in places, big slabs of sedimentary rock still adhere to its slopes.

Timpanogos Cave National Monument

Three limestone caverns, now joined with manmade passages, make up Timpanogos Cave. The interiors of the caves are finely decorated with stalactites, stalagmites, cave popcorn, helictites, and other ornaments.

The trail to the caves, which are high on the wall of American Fork Canyon, ascends from Precambrian rocks a billion years old to Mississippian rocks 330 million years old. It is an interesting climb through time, despite the absence of Ordovician, Silurian, and Devonian rock layers. The steeply tilted sedimentary rocks are part of a fault-edged anticline, cored with granite.

Caves form where weak acids in groundwater dissolve limestone. Rain and snow absorb carbon dioxide from both atmosphere and soil, and react with it to form weak carbonic acid. The acid enlarges joints, faults, and natural pore spaces in the rock, eventually opening natural caverns. The three Timpanogos caves developed along a fault zone, where water apparently found a passage through the broken rock. Fault surfaces smoothed by the movement of rock against rock remain visible in the caves today.

Much of the cavern development probably happened during Pleistocene time, with ice-age increases in snowmelt, icemelt, and precipitation. After the cave openings formed, seeping water laden with dissolved calcium carbonate dripped from cavern ceilings and sheeted down the walls to build the many cave ornaments. Stalactites grew where water dripped from cavern ceilings; stalagmites built up where drops of water splashed on cavern floors. Calcite crusts and draperies developed where water sheeted across cavern surfaces. Cave popcorn that decorates parts of the cavern developed when cavern rooms were for a time submerged.

This cave is particularly well known for its curly helictites, ornaments which develop where thin layers of fine, crystalline calcite coat cave surfaces and hold back water trying to seep into the cave. As pressure increases behind this paper-thin dam, it pushes out tiny fragments of the calcite coating. Where many successive layers of newly formed crystals are pushed out at the same spot, helictites develop, curling because the crystals of which they are made are wedge-shaped.

This is a wet cave, and cave ornamentation is an on-going process here. Every drop of water adds a miniscule bit of calcite to the ornaments. Little bumps, the beginnings of stalactites, have developed on the ceilings of manmade tunnels completed in the 1930s. Calcite lily pads grow outward into silent pools. Calcite frostwork covers older ornaments.

On Zion's Great White Throne, the Navajo sandstone breaks away along numerous vertical joints. In Zion, the upper part of the formation is white, the lower part salmon pink.

Zion National Park

At the south end of the Markagunt Plateau, the North Fork of the Virgin River carved through 2000 feet of white and pink sandstone to create magnificent Zion Canyon. The cliffs that wall the canyon shade a peaceful vale threaded by the present river.

Two Jurassic rock units, the Kayenta formation and Navajo sandstone, are responsible for much of Zion's scenic beauty. About 2000 feet thick here, and forming the massive walls and stolid towers of Zion Canyon, the Navajo sandstone accumulated on a dune-covered desert. The sweeping diagonal crossbedding of this formation, as well as its fine, rounded, frosted sand grains and its wind-formed ripplemarks, are relics of the dunes. Horizontal siltstone layers that bevel its slanting crossbedding are silt and clay blown or washed onto flat surfaces between dunes.

The Kayenta formation, below the Navajo sandstone, is made of many layers of sandstone, siltstone, and mudstone deposited on a delta or floodplain. Below its uppermost sandstone cliff it is relatively easily eroded. Because of this, the lower part of the formation plays an important role in scenery development: As it wears away, the Navajo sandstone is undermined until it breaks loose and falls away, coming to rest on steep slopes of rubble below the cliffs.

A third component important in shaping Zion Canyon is of course the North Fork of the Virgin River. Though usually quiet, the North Fork takes on fearsome proportions during heavy summer thundershowers. It then picks up sand, pebbles, and even large boulders, and with them pounds its channel and the cliffs on either side. Measurements show that at present it is cutting downward at an average rate of about an inch per century, or 600 feet per million years. Not bad, for canyon cutting. In Pleistocene time, when rainfall and snowfall were more abundant, the stream must have been torrential much of the time, and erosion doubtless was much more rapid.

Canyon widening is a continuous process, too, largely the work of slides and rockfalls. Here, again, the Kayenta formation plays a role. Watch for evidence of its role in canyon widening at Zion Canyon Narrows, near the upstream end of the scenic drive. Just at the end of the road, this formation is low on the canyon walls, at river level. A recent rockslide of broken blocks of Navajo sandstone occurred where erosion of the Kayenta formation undermined the sandstone. But just beyond the end of the paved trail, the Kayenta formation is below the surface,

Zion Canyon is walled with Navajo sandstone, with lower slopes of talus-covered Kayenta formation.
—Ray Strauss photo

not exposed to stream erosion. There, vertical Navajo sandstone cliffs plunge right down to the river, framing a narrow gorge up which neither road nor permanent trail can be built.

In most of the canyon, a line of springs marks the contact between the Navajo sandstone and the Kayenta formation. Rainfall and snowmelt soak down through the porous Navajo sandstone fairly easily. But the Kayenta formation is much less permeable, and water filtering down from above is forced to flow sideways, to emerge, where it intersects the canyon, as springs and seeps. This water speeds weathering, particularly freeze-and-thaw weathering, so some springs are in shady fern- and flower-decked overhangs of dripping rock. Similar but smaller springs mark some of the horizontal interdune siltstones in the Navajo formation, creating hanging gardens high on the canyon walls.

A number of trails zigzag up the walls of Zion Canyon or thread its intricate maze of tributary canyons. Trails to the East and West rims and to Observation Point provide especially good opportunities to look at the Navajo sandstone, its garlandlike cross-bedding, and the various shapes and forms of

Shifting winds form intricate patterns of cross-bedding in the Navajo sandstone, a product of Jurassic sand dunes. Sand is deposited on leeward dune faces.

A large arch visible from Utah 9 demonstrates the arch-forming tendencies of the Navajo sandstone. Honeycomb weathering to the left of the arch is largely the work of wind.

its erosion. Hidden Canyon Trail is as promising as its name: It leads to a little canyon formed along one of the many joints that cut the Navajo sandstone. The "trail" through the Zion Canyon Narrows (except for sand bars, you wade up the stream) is both adventure and geology lesson. Don't go in threatening weather, though. Angels Landing trail is fun if you don't mind heights.

Several of Zion's many remarkable features lie outside the canyon itself. Checkerboard Mesa, for instance, with its bare-rock slopes marked with cross-bedding and with small grooves eroded along vertical joints. Or the Kolob Canyons in the less frequented northern section of the park, site of a natural arch that rivals Rainbow Bridge for size. A petrified forest accessible by trail. Several cinder cones — small fairly recently erupted volcanoes — near the south entrance to the canyon. Many other trails — short and long — lead into other untrafficked parts of Zion.

Glossary

Alabaster: pure, dense gypsum suitable for ornamental use.

Alkali: a mixture of calcium, potassium, and sodium carbonates common in playa deposits.

Alluvial fan: sloping, fan-shaped mass of gravel and sand deposited by a stream as it issues from a mountain canyon.

Alluvial apron: an apronlike slope of gravel and sand formed by merging of alluvial fans, also called a bajada.

Alunite: a mineral used in the manufacture of alum.

Amethyst: a purple variety of quartz.

Anhydrite: a calcium sulfate mineral similar to gypsum except in containing no water.

Anticline: a fold that arches upward.

Apatite: a phosphate mineral used in manufacturing fertilizer and other products.

Aquifer: a porous, permeable rock layer from which water may be obtained.

Aragonite: a form of calcite.

Arch: a fairly large opening formed by erosion of a rock fin, not bridging a watercourse.

Artesian: water that rises above a water-bearing layer because of hydrostatic pressure.

Ash, volcanic: finely pulverized rock material thrown out by explosive volcanic eruptions. See Tuff.

Bajada: a slope of gravel and sand formed by merging of alluvial fans.

Barite: a mineral, barium sulfate, used in paints, oil well drilling muds, and paper manufacture.

Barchan dune: a crescentic sand dune with curving arms pointed downwind.

Basalt: dark gray to black volcanic rock poor in silica and rich in iron and magnesium minerals.

Base level: the theoretical limit below which a stream cannot erode.

Basin: a broad enclosed depression, commonly with no drainage to the outside.

Batholith: a mass of intrusive rock with more than 40 square miles of surface exposure.

Bed: a single layer of sedimentary rock.

Bentonite: soft, porous, light-colored rock formed by decomposition of volcanic ash.

Beryllium: a strong, lightweight, heat-resistant metal used in nuclear reactors, spacecraft nosecones, and copper alloys.

Boulder: a rounded rock fragment with a diameter greater than 10 inches.

Brachiopod: a marine shellfish having two bilaterally symmetrical shells, rare today but common as fossils.

Braided stream: a stream that divides into an interlacing network of small channels.

Breakaway zone: a zone where upper plate rocks break away and move along detachment faults.

Breccia: volcanic rock consisting of broken rock fragments imbedded in finer material such as volcanic ash.

Bryozoan: a group of tiny marine shellfish that live in corallike colonies, often found as fossils.

Butte: a steep-walled hill capped with resistant rock, smaller than a mesa.

Calcite: a common rock-forming mineral, calcium carbonate, the principal mineral in limestone and travertine.

Caldera: a basin-shaped, cliff-edged depression formed by a volcanic explosion and subsequent collapse of a volcano.

Caliche: a crusty, whitish rock that accumulates near the surface as calcite and other minerals fill pore spaces in gravel.

Carapace: a crustlike layer of metamorphosed rock surfacing a metamorphic core complex mountain, formed at depth by movement along a large detachment fault.

Catstep: a small terrace produced by slumping.

Cave popcorn: a cave deposit formed of calcium carbonate during submergence of a cave.

Chert: a dense, compact variety of quartz.

Cinder cone: a small conical volcano formed of small bits of basalt.

Cirque: a scoop-shaped, glacier-carved depression high on a mountainside, commonly at the head of a glaciated valley.

Claystone: sedimentary rock formed from clay.

Climbing dune: a sand dune formed against a cliff or steep slope.

Cobble: a rounded rock fragment with a diameter of 2.5 to 10 inches.

Columnar jointing: a pattern of vertical joints that defines parallel columns in lava and volcanic ash.

Composite volcano: see stratovolcano.

Conduit: the feeder pipe of a volcano.

Conglomerate: rock composed of rounded, water-worn fragments of older rock, typically with coarse sand between the fragments.

Continental rocks: Rocks making up the continents, lighter in color and density than those in ocean basins.

Coral: a group of marine animals that may deposit calcium carbonate in large reeflike masses.

Crater: the hollow at the top of a volcano from which volcanic material is ejected.

Creep: gradual downhill movement of soil or rock materials due to gravity.

Crinoid: a group of marine animals having jointed stems and arms, often called sea lilies.

Cross-bedding: fine diagonal layering within larger layers of sedimentary rock, due to rapid deposition by wind or water.

Crust: the outermost, cooled and hardened part of the Earth, three to 20 miles thick.

Cuesta: a ridge with a long, gentle slope of resistant caprock, and a short steep slope on eroded edges of underlying rock.

Dacite: A fine-grained volcanic rock.

Debris flow: a dense mudflow containing abundant coarse material.

Desert pavement: a veneer of tightly packed pebbles left when sand and silt are blown away.

Desert varnish: a dark, shiny surface of iron and manganese oxides found on many exposed rock surfaces in desert regions.

Detachment fault: A nearly horizontal fault, commonly formed by downward curve of normal faults, on which the brittle upper crust separates from and moves across the lower crust.

Differential erosion (or weathering): erosion (or weathering) at varying rates depending on differences in the resistance of rock layers or features within the rock.

Dike: a sheetlike intrusion of igneous rock formed when magma intrudes and cools in a vertical crack or joint.

Diorite: an intrusive igneous rock that contains more dark minerals and less quartz than granite.

Dip: the direction and degree of tilt of sedimentary layers.

Dip, primary: the original slope of a rock layer, particularly of a layer deposited on a sloping surface, as on the slopes of a volcano.

Dip slope: a slope conforming with the direction and angle of dip of sedimentary rock units, usually eroded on a resistant layer.

Dolomite: a sedimentary rock composed largely of the mineral dolomite, a carbonate of calcium and magnesium.

Dome: an anticline in which sedimentary rocks dip away in all directions.

Downthrown: the side of a fault that appears to have moved downward compared with the other side.

Earthflow: downhill movement of soil and weathered rock, similar to a debris flow.

Echinoderm: invertebrate marine animals characterized by external shells made up of many radially arranged calcite plates. Includes starfish, sea urchins, and other forms.

Elaterite: a brown, asphaltlike substance formed by changes in petroleum within the rock.

Epoch: a unit of geologic time, subdivision of a period.

Era: The largest subdivision of geologic time.

Erosion, headward: erosion at the head of a stream as rock and soil above it are undermined and washed away.

Evaporite: a mineral deposited as mineralized water evaporates, notably salt, gypsum, anhydrite, and potash.

Extrusive igneous rock: rock formed of magma that erupts onto the surface, also called volcanic rock.

Fault: a rock fracture along which displacement has occurred.

Fault block: a segment of the Earth's crust bounded on two or more sides by faults.

Fault, normal: a nearly vertical fault in which the overhanging side moves down.

Fault, reverse: a nearly vertical fault in which the overhanging side moves up.

Fault scarp: a steep slope or cliff formed by movement along a fault.

Fault, thrust: a nearly horizontal fault in which one side is pushed up and over the other side.

Fault zone: a zone of numerous small breaks that together make up many faults.

Feldspar: a group of common light-colored, rock-forming minerals containing aluminum oxides and silica.

Fill, valley: gravel, sand, and other materials washed into a valley.

Flatiron: a roughly triangular hogback on a mountain flank, shaped by erosion of a steeply inclined resistant rock layer.

Floodplain: relatively horizontal land adjacent to a river channel, with sand and gravel layers deposited by the river during floods.

Flowstone: travertine deposited in caves by water trickling across cave walls or floor.

Fold: a curve or bend in rock strata.

Foraminifera: a group of microscopic one-celled marine shellfish.

Formation: a mappable rock unit.

Fossil: remains or traces of a plant or animal preserved in rock.

Freeze-and-thaw weathering: breakdown of rock by repeated freezing and thawing of water held in pores and cracks within the rock.

Fuller's earth: very fine-grained clay used for bleaching or for absorption of impurities in oils.

Fumarole: a small vent from which gases and vapors are emitted.

Geothermal: pertaining to the Earth's heat.

Geyser: a type of hot spring that intermittently erupts jets of hot water and steam.

Glaciation: the formation and movement of glaciers or ice sheets.

Glacier: a large mass of ice formed by compaction and recrystallization of snow, which because of its weight creeps slowly downslope or outward from its center.

Glauconite: a green, earthy or granular mineral of the mica group.

Gloryhole: a cone-shaped mine resulting from deliberate blasting and caving of an ore body, with ore being removed from below.

Gneiss: banded metamorphic rock that commonly forms from granite or sandstone.

Gooseneck: an entrenched stream or river meander.

Graben: a dropped valley bounded by faults.

Granite: a coarse-grained igneous intrusive rock composed of chunky crystals of quartz and feldspar peppered with dark biotite and hornblende.

Graptolite: a tiny marine organism with a chitinous or cupshaped external skeleton arranged with other individuals in branching colonies.

Gravel: a mixture of pebbles, boulders, and sand.

Groundwater: subsurface water, as distinct from rivers, streams, seas, and lakes.

Group: a major unit of stratified rock composed of several related formations.

Gypsum: a common mineral, hydrous calcium sulfate.

Halloysite: a fine white clay made up of minute hollow tubes.

Hanging garden: a spring-nourished garden on the side of a cliff.

Head frame: the structure over a vertical mine shaft that carries the sheaves (large pulleys) over which move cables that lower and raise orebuckets and elevators.

Helictite: a small cave ornament that develops a curving, curling form.

Hematite: a common iron oxide mineral.

Hogback: a long, narrow ridge with a sharp crest, formed by erosion of layers of rock tilted approximately 45°.

Honeycomb weathering: wearing of rock into many cell-like hollows or tafoni.

Hoodoo: a fantastic, often grotesque column or pinnacle of rock.

Horst: an uplifted block of crust bounded on its long sides by faults.

Hydrothermal alteration: alteration by hot groundwater.

Igneous rock: rock formed from molten magma.

Interdune deposits: fine-grained silt or mud deposited in flat areas between dunes.

Intrusion: a body of igneous rock intruded while molten into older rock.

Intrusive igneous rock: igneous rock created by subsurface cooling of molten magma.

Inverted valley: a lava-filled valley whose bordering ridges have eroded away leaving the lava higher than its surroundings.

Jasper: a dense, opaque, red or yellow variety of chert.

Joint: a rock fracture along which no significant movement has taken place.

Kettle: a depression in a glacial deposit, formed as a large chunk of ice melted.

Laccolith: a lenslike intrusion that spreads between rock layers, doming those above it.

Landslide: a general term for downhill movement of rock and/or soil — due to the pull of gravity. Includes rockfalls, mudflows, debris flows, slumps, etc.

Lava: igneous rock created by cooling of molten magma on the Earth's surface or under the sea.

Lava dome: a dome-shaped body of volcanic rock formed from very thick magma.

Lichen: a plant community consisting of a fungus and an alga, appearing as a flat crust on a rock surface.

Limestone: a sedimentary rock consisting of calcium carbonate, usually formed from the shells of marine animals and calcium-secreting plants.

Limonite: a yellow-brown hydrous iron oxide mineral.

Lithosphere: the solid outer portion of the Earth, consisting of the crust and upper mantle.

Magma: molten rock.

Magma chamber: An underground reservoir of magma from which volcanic materials are derived.

Magnetic reversal: a shift in the Earth's magnetic field, with the north and south magnetic poles switching places.

Magnetite: a heavy black magnetic iron mineral common in sand and in igneous rocks.

Malachite: a bright green copper mineral.

Mantle: The zone between the Earth's core and crust.

Marble: metamorphic rock consisting of recrystallized limestone.

Meander: a looplike bend in a river.

Meander, entrenched: a looplike river bend carved deeply into underlying rock.

Member: a subdivision of a formation.

Mesa: a flat-topped hill or mountain capped with a resistant rock layer and edged with steep cliffs or slopes.

Metamorphic core complex: a dome-shaped mountain formed as movement on a detachment fault unroofs part of the lower crust, which then rises because of its own buoyancy.

Metamorphic rocks: rocks formed from older rocks by great heat and pressure or by chemical changes.

Mica: a group of shiny minerals easily separated into thin, shiny flakes.

Mid-ocean ridge: a volcanic ridge running across ocean basins, the site of sea floor spreading.

Mineral: a naturally occurring substance with a characteristic chemical composition and usually with typical color, texture, and crystal form.

Monocline: a fold in stratified rock in which all the strata dip in the same direction.

Monzonite: an intrusive igneous rock with a high proportion of feldspar minerals and very little quartz. Quartz monzonite contains more quartz.

Moraine: a mound, ridge, or surface layer of unsorted glacier-deposited rock material.

Moraine, ground: rock material deposited over a wide surface as a glacier melts.

Mudcrack: a shrinkage crack in drying mud.

Mudflow: a flowing mass of fine mud.

Mudstone: sedimentary rock formed from mud.

Muscovite: white mica.

Natural bridge: a rock span formed by erosion across a ravine or valley, usually where a stream or other watercourse abandoned a meander bend.

Obsidian: black volcanic glass.

Oceanic rocks: rocks making up the oceanic crust, usually darker and denser than those of the continents.

Oil shale: fine-grained rock containing a waxy substance called kerogen, from which oil or gas can be distilled.

Olivine: an olive green or yellow mineral common in basalt.

Oolitic: made up of minute spheres.

Orogeny: mountain-building

Oxbow lake: a crescent-shaped lake left when a stream or river cuts through the neck of a meander and isolates its original channel.

Parabolic dune: a horseshoe-shaped sand dune in which long arms partly stabilized by vegetation point in a windward direction.

Pebble: a rock fragment, commonly rounded, 0.2 to 2.5 inches in diameter.

Pediment: a gently inclined erosion surface carved in bedrock at the base of a mountain range.

Period: a subdivision of geologic time shorter than an era, longer than an epoch.

Phyllite: a shiny, scaly, fine-grained metamorphic rock.

Placer gold: gold found as free particles or flakes in gravel and sand.

Plate: a block of the Earth's crust, separated from other blocks by mid-ocean ridges, trenches, and collision zones.

Playa: a flat-floored dry lake bottom in an undrained desert basin.

Playa lake: a shallow intermittent lake formed on a playa after rains.

Potash: potassium carbonate used in the manufacture of fertilizer.

Pothole: a circular hollow dissolved in a rock surface by standing water or excavated by whirlpools using stone and sand tools.

Pumice: pale, frothy volcanic rock, often light enough to float on water.

Quartz: a hard, glassy, rock-forming mineral composed of crystalline silica, silicon dioxide.

Quartzite: a metamorphic rock formed of tightly cemented quartz sand grains.

Racetrack valley: a circular or oval valley formed by erosion of soft rock layers around a dome of harder rock.

Radiometric dating: dating of rocks by measurement of radioactive minerals and their decay products.

Redbeds: red sedimentary rocks, usually siltstone and mudstone.

Rhyolite: a light gray volcanic rock with large quartz and feldspar crystals in a finer groundmass, the extrusive equivalent of granite.

Ring dike: an arcuate or circular dike usually formed around the site of a former volcano.

Salt anticline: an anticline formed by concentration and upward movement of thick salt deposits.

Sandstone: sedimentary rock formed from sand.

Scarp: a low cliff caused by movement along a fault or by landslide movement.

Schist: metamorphic rock with parallel orientation of mica grains that causes it to break easily along parallel surfaces.

Scoria: bubble-filled lava.

Sedimentary rock: rock formed from particles of other rock transported and deposited by water, wind, or ice.

Seismic: pertaining to earthquakes or manmade earth vibrations.

Selenite: a silky, fibrous variety of gypsum common in veins in sedimentary rocks.

Shale: a fine-grained sedimentary rock formed by consolidation of clay, silt, or mud, characterized by breaking into flat sheets.

Sheave: a large pulleylike wheel by which cables are lowered into a mine shaft.

Silicic: containing more than 65% of silica (quartz) or silicate minerals.

Silicified: penetrated and preserved by silica.

Siltstone: rock formed from particles smaller than those of sand and larger than those of clay.

Sill: a thin body of igneous rock intruded between horizontal rock layers. Or the lowest point in land surrounding a lake, where the lake can overflow.

Slate: a metamorphic rock formed from shale, that splits along surfaces other than bedding planes.

Slickenside: a grooved and polished rock surface along a fault.

Slump: a landslide characterized by movement along a curved slip surface, with backward tilting of the slumped mass.

Slurry: a wet, highly mobile mixture of rock materials and water.

Stalactite: a dripstone icicle that hangs from the roof of a limestone cavern.

Stalagmite: a dripstone pedestal that rises from the floor of a limestone cavern.

Strata: layers of sedimentary rock. Singular is stratum.

Stratified: formed in layers, as sedimentary rock.

Stratigraphic: pertaining to layered rocks.

Stratovolcano: a cone-shaped volcano built of alternating layers of lava and volcanic ash. Also called a composite volcano.

Stromatolites: variously shaped calcareous fossils thought to be the remains of algae.

Subduction: the drawing downward of an oceanic plate as a continental plate overrides it.

Syncline: a troughlike downward fold in sedimentary rocks.

Tafoni: small hollows eroded in rock by wind and rain.

Talus: fallen rock fragments at the base of a cliff.

Tar sand: sand or sandstone saturated with tar or very thick oil.

Tepee rock: a cone-shaped rock eroded in volcanic ash.

Terrace: a relatively level bench bordering a river valley, representing an earlier river floodplain or lake shoreline.

Thrust belt: a region where large thrust faults bring older rocks over younger rocks.

Trap: a structure in rock layers that tends to capture oil.

Transverse dune: a sand dune in the form of a long, often sinuous ridge at right angles to the prevailing wind direction.

Trench: a long, deep oceanic depression formed when an oceanic plate is pulled downward and overridden by a continental plate.

Trilobite: an extinct marine arthropod characterized by a three-lobed, oval-shaped body.

Tufa: a porous sedimentary rock formed of calcium carbonate deposited around a spring or along lake shores.

Tuff: a rock formed of compacted volcanic ash and cinders.

Tuff, ashfall: tuff formed from volcanic ash falling from an overhead ash cloud.

Tuff, ashflow: tuff formed from very hot volcanic ash that flows down the flank of a volcano and welds itself together as it stops.

Type locality: the locality at which a named rock unit was originally studied and described.

Unconformity: a surface of erosion or non-deposition that separates younger strata from older rocks.

Valley fill: sand, gravel, and volcanic material filling a valley.

Vein: thin, sheetlike mineral material filling a joint or fault.

Vent: a volcanic opening through which lava, cinders, volcanic ash, or volcanic gases escape.

Vesicle: a bubble cavity in hardened lava.

Volcanic ash: see tuff.

Volcanic rock: rock formed by cooling of molten magma on or very near the surface. Also called extrusive igneous rock.

Volcano: a mountain or hill built by escape of magma to the Earth's surface.

Volcano, shield: a dome-shaped volcano formed by moderately fluid lava.

Warm spring: a thermal spring with a temperature below that of the human body.

Wash: a western term for a broad, shallow, sand or gravel streambed, dry much of the time.

Waterpocket: a depression where water may gather, particularly a pocket-shaped depression in solid rock.

Water table: the surface below which rocks and soil are saturated with groundwater.

Weathering: changes in rock due to exposure to the atmosphere.

Window: a small opening high up through a rock fin.

Index

Check for our books at your local bookstore. Most stores will be happy to order any which they do not stock. We encourage you to patronize your local bookstore. Or order directly from us, either by mail, using the enclosed order form or our toll-free number, 1-800-234-5308, and putting your order on your Mastercard or Visa charge card. We will gladly send you a complete catalog upon request.

Questions? Call us toll-free, 1-800-234-5308.

Some other geology titles of interest:

Title	Price
____ROADSIDE GEOLOGY OF ALASKA	12.95
____ROADSIDE GEOLOGY OF ARIZONA	12.95
____ROADSIDE GEOLOGY OF COLORADO	11.95
____ROADSIDE GEOLOGY OF IDAHO	15.00
____ROADSIDE GEOLOGY OF MONTANA	14.95
____ROADSIDE GEOLOGY OF NEW MEXICO	11.95
____ROADSIDE GEOLOGY OF NEW YORK	12.95
____ROADSIDE GEOLOGY OF NORTHERN CALIFORNIA	11.95
____ROADSIDE GEOLOGY OF OREGON	11.95
____ROADSIDE GEOLOGY OF PENNSYLVANIA	12.95
____ROADSIDE GEOLOGY OF TEXAS	15.95
____ROADSIDE GEOLOGY OF VERMONT & NEW HAMPSHIRE	9.95
____ROADSIDE GEOLOGY OF VIRGINIA	12.00
____ROADSIDE GEOLOGY OF WASHINGTON	12.95
____ROADSIDE GEOLOGY OF WYOMING	11.95
____ROADSIDE GEOLOGY OF THE YELLOWSTONE COUNTRY	9.95
____AGENTS OF CHAOS	12.95
____FIRE MOUNTAINS OF THE WEST	15.95
____IMPRINTS OF TIME: THE ART OF GEOLOGY	19.95

Please include $2.00 per order to cover postage and handling.

Please send

Name

Address

City

☐ Paym

Bill my: ☐

Card No.

Signature

M